The Chosen Species

For Lourdes
For Chon

JUAN LUIS ARSUAGA AND IGNACIO MARTÍNEZ

Translated by Rachel Gomme   Illustrations by Mauricio Antón

THE CHOSEN
SPECIES

The Long March of Human Evolution

Blackwell
Publishing

Spanish edition © 1998 by Juan Luis Arsuaga Ferreras, Ignacio Martínez Mendizábal, and Ediciones Temas de Hoy, Ltd. (T.H.)
Illustrations © 1998 by Mauricio Antón
English translation © 2006 by Blackwell Publishing, Ltd.

BLACKWELL PUBLISHING
350 Main Street, Malden, MA 02148-5020, USA
9600 Garsington Road, Oxford OX4 2DF, UK
550 Swanston Street, Carlton, Victoria 3053, Australia

Originally published in 1998 by Temas de Hoy as *La especie elegida*
English edition published 2006 by Blackwell Publishing
Translated into English by Rachel Gomme

1   2006

Library of Congress Cataloging-in-Publication Data

Arsuaga, Juan Luis de.
[Especie elegida. English]
The chosen species : the long march of human evolution / Juan Luis Arsuaga, Ignacio Martínez ; illustrations by Mauricio Antón ; [translated into English by Rachel Gomme].
   p. cm.
"Originally published in 1997 by Temas de Hoy as La especie elegida"—T.p. verso.
Includes bibliographical references and index.
ISBN-13: 978-1-4051-1532-2 (hard cover : alk. paper)
ISBN-10: 1-4051-1532-7 (hard cover : alk. paper)
ISBN-13: 978-1-4051-1533-9 (pbk. : alk. paper)
ISBN-10: 1-4051-1533-5 (pbk. : alk. paper)
1. Fossil hominids. 2. Human evolution. 3. Primates—Evolution. 4. Anthropology, Prehistoric. I. Martínez, Ignacio. II. Antón, Mauricio. III. Title.

GN282.A69813 2006
599.93′8—dc22

2004027275

A catalogue record for this title is available from the British Library.

Set in 10/12.5pt Galliard
by SPI Publisher Services, Pondicherry, India
Printed and bound in the United Kingdom
by TJ International Ltd, Padstow, Cornwall

The publisher's policy is to use permanent paper from mills that operate a sustainable forestry policy, and which has been manufactured from pulp processed using acid-free and elementary chlorine-free practices. Furthermore, the publisher ensures that the text paper and cover board used have met acceptable environmental accreditation standards.

For further information on
Blackwell Publishing, visit our website:
www.blackwellpublishing.com

Contents

Acknowledgments   xi

Introduction: Prehistory   1
Little Lucy   1
Intrepid Paleontologists   3

Part I   Children of Africa   7

1 Basic Principles of Evolutionary Theory   9
      The Inheritance of Acquired Characteristics   9
      Natural Selection   10
      The Staircase of Progress   13

2 We the Primates   17
      The Ecological Definition and Diversity of Primates   17
      The Classification of Primates   20
      Hominoids, Apes of Our Own Branch   22
      History of the Primates   27

3 Climate and Evolution   35
      The Origin of Species   35
      Climate Changes over the Last Few Million Years   36
      Paleotemperature Scales   38
      Factors in Climate Change   39

Contents

Milankovitch Cycles   40
A Climatic Model for Equatorial Africa   43
The Controversial Gas   45
The End of Paradise   46

4 The Origin of Humanity   49
   Molecular Clocks   49
   The First Fossil Hominids   51
   Change of Habitat   53
   East Side Story   57
   Dating Fossils   59
   The Taung Child   61
   Distinguishing Marks   62

5 The Bipedal Chimpanzee   66
   The Great Step   66
   The Laetoli Footprints   74
   The Mystery of Mysteries of Human Evolution   76
   Portrait of the Entire Body of an Australopithecine   78

6 *Paranthropus* – Hominids of the Open Plains   87
   The Emergence and Distribution of Paranthropus   87
   The Specialist   94

7 A New Kind of Hominid   97
   The First Humans   97
   The Stone Cutters   98
   The Diversification of Homo   102
   Ready for the Great Leap   105
   Family Relationships   107
   The Science of Relationships   110
   The Hominid Tree   112

8 The Evolution of the Brain   116
   The Organ of Intelligence   116
   Large and Small Brains   117
   World Champions of Encephalization   120
   Weighing Ghosts   120
   The Brain Size of Fossil Hominids   123
   Surface Area of the Brain   124
   The Size of the Intellect   127

9 Teeth, Guts, Hands, Brain   129
    Types of Diet   129
    Carnivorous and Herbivorous Mammals   132
    The Teeth of Primates   134
    The Teeth of the First Hominids   136
    Size of the Molars and Shape of the Hand   138
    Guts and Brain   142

10 Development   145
    The Rhythm of the Molars   145
    Birth and the Newborn   146
    Childhood and Adolescence   152

11 Social Intelligence   155
    The Unexciting Sex Life of the Female Orangutan   155
    Behavior as Adaptation   156
    Comparative Sociobiology of Hominoids   158
    Natural Selection and Sexual Selection   160
    Bipedal and Monogamous from the Beginning?   162
    Brain Size and Size of Social Group   169

Summary   171

Part II A New Home   177

12 New Locations for Human Evolution   179
    Homo erectus and the Settlement of Asia   179
    The First Europeans   182
    Gran Dolina and the First Europeans   184
    Prehistoric Cannibalism   186
    Homo antecessor   187
    Human Evolution in Europe during the Middle
    Pleistocene   190
    The Pit of Bones   192

13 The Neanderthals   197
    The Way They Were   197
    Life and Death among the Neanderthals   205
    The Beginning and End of the Neanderthals   208

Contents

14 The Origins of Modern Humanity: The Fossil Evidence 212
   Neanderthals and Modern Humans 212
   Two Intelligent Human Species 216
   The Levant: A Crossroads 219
   To the Easternmost Edge of Asia 222
   The African Origin of Homo sapiens 223

15 The Origins of Modern Humanity: The Genetic
   Evidence 226
   A Brilliant Idea 226
   The Molecules of Inheritance 228
   Black Eve 229
   An Adam for Eve 231
   The Other Chromosomes 232
   Pleistocene Park 233
   Fossils and Molecules 235
   Standards of Beauty 237

16 The Origins of Human Language 239
   King Solomon's Ring 239
   Language and Brain 241
   The Choking Primate 243
   The Production of Speech 244
   The Fossils Speak 247
   Group Selection and the Extinction of the Neanderthals 249
   The Crooked Lines of Natural Selection 253

17 The Meaning of Evolution 255
   The Action Replay of Life 255
   Organization and Chaos 261

Epilogue 264
The Never-ending Story 264

Bibliography 268

Index 275

For now consider, like the zoologists and the anatomists, that man has more of the ape than of the angel, and that he has few grounds for vanity or conceit.

Santiago Ramón y Cajal

Acknowledgments

The authors are indebted to many people whose contributions helped to improve this book. First and foremost to Jesús Arsuaga, who devoted much time to investigating and summarizing what is known today about climate changes and their influence on human evolution. The majority of the chapter on climate is his work. Manuel Martín-Loeches made very interesting suggestions on the subject of the brain. Elena Benavente and Jesús Pérez-Gil reviewed the genetic evidence. We, and we hope the book also, benefited from our "philosophical" conversations with Fernando Palacios. And from our co-researchers José Miguel Carretero, Nuria García, Ana Gracia, and Carlos Lorenzo, we consistently received generous support and valuable ideas.

Introduction: Prehistory

The main conclusion arrived at in this work, namely, that man is descended from some lowly organized form, will, I regret to think, be highly distasteful to many.

Charles Darwin, *The Descent of Man*

Little Lucy

Little Lucy trudged her way laboriously through the African savanna. As a result of continuous effort over generations, she walked on two legs rather than four like her ancestors. With her child weighing heavy in her arms, she felt herself weakening as she drew nearer to the clump of thorny acacias which could be glimpsed on the horizon, under the torrid tropical sun. Small as she was – little more than a meter in height, and weighing less than 30 kilograms – it was only by her wits that she had evaded powerful predators. She had no stone tools. A million years had passed since her ancestors, the first hominids, had decided to leave the protection of the forest and move out into the savanna which continued to expand, as a result of the great climate change then taking place. Her relatives, the ancestors of chimpanzees and gorillas, had preferred the safety of the forest and would remain there forever. For destiny belonged to the bold, those who defied the danger of open environments. One day these bolder beings would evolve, their brains and their intelligence would develop, they would manufacture all kinds of tools, they would discover fire and would banish forever the lion, the leopard, and the hyena.

1

Eventually, they would conquer the world. All of this depended on Lucy and her baby surviving and joining the little group of australopithecines waiting in the stand of trees, who represented the future of humanity. Lucy literally carried our future on her two legs.

But no, dear reader, this is not one of *those* books; we have not allowed our imagination to run away with us. This is not a fanciful tale about human evolution. Its aim is not to keep you in suspense, holding your breath over the vicissitudes of our ancestors. After all, the ending of the story of human evolution is well known. Lucy survived, if you like to think of it those terms. Ultimately we are all descended from many Lucys.

Misia Landau has drawn attention to the narrative structure of histories of human evolution, the rhetoric which surrounds them and its similarity to mythological and religious literature. This is evident in the biblical references in some of the names given to scientific hypotheses about our origin, such as the Black Eve hypothesis, the Noah's Ark hypothesis, or even the title of this book. In fact in other cases the only aim of such names is to catch the reader's attention with familiar-sounding titles, like the Out of Africa hypothesis, or the East Side Story hypothesis. But the really important issue is not the name given to these hypotheses, but that they can be judged against the facts, modified and even rejected if they are not compatible with them: this is what makes them scientific hypotheses rather than simply opinions or fantasies. Only dogma is immutable.

For however her story is told, Lucy is much more than a myth. Lucy is real. The man who discovered her, Donald Johanson, gave this name in 1974 to an extremely well-preserved skeleton of a female hominid who lived in what is now known as Ethiopia some three million years ago. And she really was small. She no longer lived in the enclosed, humid forests, but rather in more open, arid spaces: although this environment increased the risk of falling victim to predators, it offered new plant resources for food. Lucy could not speak as we do; her brain was not much bigger than that of a chimpanzee, and she had no stone tools, but she was bipedal. All the other elements of our tale – the split between the lines leading to chimpanzees and to our species, when and where it took place, the climate change which occurred and the reduction in the tropical forest which resulted from it – all of this is based on scientific data. Our only artistic license is in relating the life-story of a particular hominid individual.

Nevertheless, there are elements in the structure of this story which remain hidden from view, but which have profound implications for correct understanding not only of human evolution, but of evolution in general. It is therefore worth pausing a moment to analyze these narratives.

First, the tale is evolutionist: it accepts that our species originated through evolution from other species, thus forming a long, continuous chain through time. However, the structure of the story as we presented it here is not Darwinist, because it contains a subtle acknowledgment of the active role of living organisms in their own evolution, setting them as protagonists confronted by a changing environment. What is more, in the first paragraph of the story the development of an erect walking position is attributed to continuous effort and exercise. Darwin, on the contrary, believed that organisms are passive subjects in evolution: they constitute the raw material which natural selection molds, giving form to different and changing species over time, and the activities undertaken by individuals during their lifetime have not the slightest effect on the anatomical structures and organs which they will pass on to their offspring.

Finally, and most importantly of all, the story with which we open this book does not present evolution as a process directed by either internal or external forces which guide it, according to a preestablished plan or design, towards its culmination in the human being. Lucy could easily die and her lineage be lost; the entire species could have become extinct, and we would not be here today. In other words, the implication of this tale is that we are not the inevitable consequence of the evolutionary process; we were at the mercy of fate.

The majority of the people with whom we have discussed human evolution outside of our professional sphere had no problem with acknowledging the evolutionary origin of our species, but they were convinced nevertheless that we are the "most evolved" species, the culmination of the entire evolutionary process – in short, the "chosen species." Since this appears to be a widely held conviction, and since in order to discuss it we need to understand how evolution occurs, what it consists of, and where it is headed (if indeed it is headed anywhere), we shall devote the first and the last chapters of this book to this crucial question.

Intrepid Paleontologists

This is also not a book about daring paleoanthropologists and their adventures discovering human fossils, although we could tell some of these tales from first-hand experience.

Scientific research is always an intellectual adventure, one which poses challenges, attempts to reach new horizons of knowledge, and must overcome numerous obstacles with large doses of ingenuity and effort. But paleontology is doubly an adventure, because the objects of study must be sought out in the field itself, in nature. When scientists speak of a discovery, they are generally referring to the discovery of some law or property, or the solution to a complex problem which may perhaps be expressed in terms of a formula. In paleoanthropology, in addition to this type of discovery, a new find may, in exceptional cases, take the most solid and material of forms. It might consist of the fossilized remains of one of our remote ancestors. Paleontologists are the only scientists who are able to travel far back in time and, in the case of paleoanthropology (the study of ancient humans), transport us to any given moment in the history of our origins. We hope that this book will communicate to the reader our passionate enthusiasm for the search for our ancestors, without the need to describe our emotions at those unforgettable moments of discovery of fossil humans – emotions which we shared with our companions in toil and which, to tell the truth, we could not express in words.

Every time we gave a lecture we could sense the interest which the subject of human evolution awakens in the most varied of audiences. But at the end of the discussions those present were too shy to formulate the questions which occurred to them, because they seemed too basic, unworthy of being put to a professional paleontologist. People do not realize that the questions everyone wonders about are the same as those the scientist tries to solve, and that they are often the most difficult to answer. How do we know the age of fossils? Where and when did we emerge? Have we been "murderous apes" from the beginning of our history? What came first, a being that walked on two legs or one that was intelligent? Were our ancestors monogamous? Why does childbirth hurt? How long did childhood last among primitive hominids? What did they eat? How tall were they? When did humans first begin to speak? Are we the hominid species with the biggest brain? This book is designed to respond to these questions. But in order to answer them, they need to be posed in the right way, and placed in the context of human evolution.

To some extent, the work of the paleoanthropologist resembles detective work. Like the detective, the paleoanthropologist arrives at the scene of the "crime" after it has already happened. On the basis of indirect information, he must reconstruct the sequence of events and – the more difficult part – find logical explanations which enable us to understand what has happened; both the detective and the paleoanthropologist must give an account of the "how" and the "why" of the events that took place.

Good detective novels provide the reader with all the clues and the detective's deductions to elucidate the case. It is very irritating to read to the end of the novel only to find that the solution depends on evidence which only the detective knew about and which has been concealed from the reader until that point. But what is really unforgivable in a detective novel is if the solution to the case is not explained in a satisfactory way, since the most interesting thing is not "whodunit," but how it was worked out. This is because the detective novel appeals to the reader's intelligence. But if the reader is to have all the information at his disposal and then appreciate the detective's shrewd work, he needs to be present at the interviews with all the witnesses, to observe the scene of the crime at his leisure, to investigate suspects' past history, study the results of laboratory analysis, and take time to reflect in order to try and fit all the pieces of the puzzle together.

Well, this book also appeals to the intelligence of our readers, and we have therefore tried to ensure that you will find in its pages all the facts and arguments on which our conclusions are based. To this end, the book follows the chronological axis of the evolution of hominids, and tackles the key questions at different points. As in a detective novel, the reader can skip straight to the end to find the solution, but this will mean missing most of the plot; moreover, the file remains "open" on many of the questions of human evolution.

Two quotations cited in Fernando Trueba's *Diccionario del cine* [Dictionary of Cinema] influenced the conception of this book, we hope to the good. One of these will be discussed in the epilogue. The other is apposite here, because it was a great solace when we were attempting to make some of the really very complicated scientific problems easily comprehensible for a general audience. Under the keyword "simplicity" in Trueba's book, he quotes the following phrase from Albert Einstein: "Everything must be made as simple as possible, but no simpler."

I

Children of Africa

Basic Principles of Evolutionary Theory

There seems to be no more design in the variability of organic beings and in the action of natural selection, than in the course which the wind blows.

Charles Darwin, *Autobiography*

The Inheritance of Acquired Characteristics

It would appear logical that the evolution of species is directly related to the life habits of the individuals which make up those species. If our ancestors lived in and moved through trees, over the course of their individual lives they would naturally learn to jump and climb. We might assume that this would result in changes in their physical constitution which would be inherited by their offspring, who would in their turn refine the modifications and pass them on, with improvements, to the succeeding generation. If at some point a group of apes began to come down from trees and adopted the habit of walking on the ground on two feet, the exercise of this activity would make things easier for their descendants. The latter, by continuing in this new form of locomotion, would gradually, through use and over generations, modify the anatomical structures necessary for walking upright.

This was how Jean-Baptiste de Lamarck (1744–1829) understood evolution at the beginning of the nineteenth century. His theory was based on the principle that the transformations occurring in individuals over their lifetime, through the use or lack of use of organs and structures, would be transmitted to their offspring. In his best-known example, he suggested that giraffes had acquired their long necks by stretching them, over generations, to reach the leaves of trees.

Although this explanation seems quite reasonable, unfortunately for Lamarck the natural world is not governed by human logic. What we have discovered of the laws of heredity, from Gregor Mendel (1822–84) up to now, leads us to reject Lamarckian theory. Whatever we do, we cannot modify the genes our children will inherit. However much we swim during our own lifetime, they will have to start from square one. The laws of biological inheritance are not like human laws.

Natural Selection

Years later, an alternative to Lamarck from within the evolutionist camp was put forward in the work of Charles Darwin (1809–82) and Alfred Russell Wallace (1823–1913). These scientists took the view that individuals had no active role in evolution. The resources of the environment being limited, only a few of those that are born will succeed in reproducing. Given that all the individuals in a given species are genetically different (except for monozygotic [identical] twins, developed from a single fertilized ovum), the competition which ensues will see some at an advantage and others at a disadvantage because of their genes: this is how selection occurs.

At the end of his life when, in 1876, he was writing a brief autobiography for his children, Darwin continued to be amazed at how innumerable organisms of all types were so wonderfully adapted to their life habits (what today we would call their ecological niche). By way of example he cited the woodpecker's adaptation to enable it to climb trees; or seeds which have developed parachute-like down to help them to disperse on the wind, or hooks which allow them to attach themselves to animals' fur. He explained this in terms not of the will of the organisms themselves, but of *natural selection* which, he suggested, preserves favorable variations and destroys unfavorable ones.

Although Darwin discovered early on that the key to evolution lay in a selection similar to that which had been practiced on domestic animals and plants since Neolithic times, it was not until October 3, 1838 that he came to understand how this principle could be applied to organisms living in the wild. On that date Darwin read an essay by the economist and demographer Thomas Robert Malthus (1776–1834), which stated that if they were not checked, human populations would tend to grow in geometric progression, outstripping the increase in resources.

There is a kind of sentimental ecological thinking which has become popular among city-dwellers, and which hampers correct understanding of natural selection. Many people believe that animals, when they are not persecuted by humans, lead a pleasant, risk-free life in nature. If this were the case, why would living beings have to adapt to become more efficient?

On the contrary, simple arithmetic shows us that, in the natural world, the life of animals is far from smooth. In demographically stable populations – in other words, populations which are not growing, which includes all populations in normal circumstances over the long term, although there may be short-term fluctuations – each reproducing pair is replaced by another two individuals in the succeeding generation. Nevertheless, in favorable conditions, a zebra living on the African plains gives birth to a foal each year from the age of 4, over a period of 15 or more years; a gazelle gives birth to one kid every 6 months from the age of 2. It is obvious that the majority of those born will not reach adulthood or reproduce. The predators fare no better: lions begin to reproduce at the age of 4 (they can easily live up to 15 years in the wild), and have 2 or 3 cubs every 20 or 30 months; leopards living in the wild reproduce from the age of 2 to the age of 12, producing from 1 to 3 cubs at intervals of about 25 months.

The same reasoning applies to primates, and, over the period of their evolution, to humans as well, although our situation has changed in recent times, since the infant mortality rate has fallen so much that without birth control we see an explosion of the population. The proportion of fertilized eggs which do not become reproducing adults is almost one hundred percent in the majority of aquatic vertebrates (amphibians, fish), and in almost all invertebrates. The upshot of these simple figures is that individuals in the different species are constantly under the threat of death, and that consequently small genetic advantages may be crucial to reaching adulthood and reproducing, or to continuing to reproduce. This is what Darwin meant when he spoke of the struggle for existence, which is not necessarily a bloody battle: plants and herbivores also compete with one another.

Unlike artificial selection, performed by humans with animals and plants and gradually enhancing specific characteristics in order to increase productivity, natural selection pursues no particular goal. Moreover, no genetic variant is better than another in an absolute sense; everything depends on the environmental circumstances. What is favorable at one moment may be unfavorable at another. In addition, through the phenomenon known as *mutation*, from time to time individuals with new variations are born, but the habits and needs of individuals do not in any way determine the direction the mutations will take. Nevertheless, these mutations represent an inexhaustible source of new variations on which natural selection can act, modifying species over time and pushing forward their evolution. Mutations do not in and of themselves produce new species; rather they increase the variation of existing species.

Chance also plays an important part in evolution, for example when a few individuals fortuitously (by pure good luck) survive an ecological catastrophe which decimates their species, or when a few individuals are passively transported by natural forces (wind, rivers, or sea currents) to found a new population. The characteristics of these randomly selected individuals might not be the most common in the original population, but they will nevertheless form the starting point for subsequent evolution. Sometimes a catastrophe on a larger scale may eliminate one or more perfectly well-adapted species at a stroke, as we shall see below.

In general terms, this is the basic theory commonly accepted by the scientific world since the 1940s. It is known as *neo-Darwinism*, because it integrates, in a modern synthesis, the ideas of Darwin and advances in genetics and other areas of biology, including the study of fossils, or paleontology.

Within the evolutionist camp there are those who contest this vision of a gradual evolution, proceeding by small steps like those patiently taken in the artificial selection of domestic animal breeds. Authors such as Stephen Jay Gould and Niles Eldredge believe that evolution advances in great strides, or even leaps. In other words, the great evolutionary innovations, the appearance of the large groups of organisms such as birds or vertebrates, is due not to the gradual accumulation of small changes, but to radical transformations.

In fact, living organisms are mechanisms so complex, and at the same time so perfectly well tuned, that it is difficult to understand how mutants radically different from their progenitors and yet capable of surviving – what Richard Goldschmidt (1878–1958) called "hopeful monsters" – can appear. Various explanations of the viability of these "monsters with a future" have been put forward. These include changes in the developmental process which, acting on both the pathways and rhythms of development, would result in adults surprisingly different from their parents – for example, having some exaggerated characteristics, or conversely, appearing infantile in certain aspects.

On the other hand, it is also difficult to understand how natural selection can detect minute changes in order to favor them. The time factor has been invoked to support the model of slow evolution, suggesting that infinitesimal variations would give those who carried them the tiniest of advantages which would only become dominant over many generations. And there is no shortage of time in paleontology. After all, life has existed on earth for more than 3,500 million years. It is around this tension between two extremes, gradual evolution or evolution by leaps, that current debate within evolutionary theory is focused.

The Staircase of Progress

Let us broaden our focus, moving from evolution at species level, or *microevolution* (which is measured in hundreds of thousands of years), to look at evolution in its wider context, or *macroevolution*, which involves whole groups of organisms comprising many species, and operates on a time scale of millions of years. If we look back at the course of our own evolution, do we not see a tendency to more and more complex and intelligent forms, culminating in our species? Are we not the predictable result of evolution? And is our species not, as is commonly held, the most evolved of all? And now that we have arrived at this point, is human evolution complete or will it continue toward even more intelligent, even more perfect forms?

This theory of evolution as a staircase of progress which leads to the species *Homo sapiens* is deep-rooted in our society, and no less so in scientific and academic circles, albeit subconsciously. For many years one of the authors taught a university course in human paleontology, and readers may already have guessed where this course was timetabled in the undergraduate program: in the last half of the last semester of the final year. Paleontology books which discuss evolution also place the evolution of our species in the last chapter, after that of unicellular organisms and, in strict order, invertebrates, amphibians, reptiles, birds, and the other mammals. Under these circumstances it is difficult for anyone to escape the idea that evolution ends with us, perhaps for ever. Even terms such as lower and higher vertebrates, or lower and higher primates, are still used (naturally, in both cases we are classified among the higher categories).

We doubt whether there is any teaching program or book which begins with the first forms of life on the planet and ends with sea urchins or insects (plants are always relegated to the margins), with mammals, primates, and humans lost among the intervening chapters or lectures. Does this mean that vertebrates are "better" than invertebrates, that mammals are "better" than reptiles, that primates are "better" than other mammals, and finally, that humans are "better" than chimpanzees and gorillas?

According to Darwin, evolution has no goal, it follows no preconceived design; it is simply opportunist, and is not directed toward any idea of perfection. To put it another way, all species (including our own) are equally perfect, each one marvelously adapted to its life habits through natural selection. In other words, unlike artificial selection which the farmer or stockbreeder carries out with a particular aim, natural selection has no objective. Although in common parlance (as well as in politics and business) the word "evolution" signifies change for the better, in Darwinian terms "evolution" means simply change, nothing more.

Among the mammals, humans are distinguished as tailless, bipedal primates with a large brain, but otherwise we have few original features. We still retain five digits on our hands and feet, while horses support themselves on the third phalange of their single digit. We show nothing like the transformations which bats or dolphins underwent, to evolve from their quadruped ancestors. Are we more evolved, in the sense of being more changed, than they are? Granted, a geranium cannot write a book – that is one of our specialties – but with the aid of light it can synthesize organic matter from mineral salts, water, and carbon dioxide; there is no doubt that the geranium has a well-equipped laboratory, and it is difficult to see it as an "inferior" being.

But those who prefer to see evolution as an arrow which has been pointing toward us since the beginning will have to answer the question of what obscure internal forces could be driving evolution in the right direction, independent of what happens around it. Or are we really dealing with forces outside of the natural world? In this case we move outside of the sphere of science, which is the terrain of this book and its authors. The object of science is to explain natural phenomena, like the existence of our species (and all the others), in terms of natural causes.

So, to return to the terrain of science, Lamarck believed in the idea of progress in evolution. However, the mechanism he proposed as the motor driving evolution forward was *adaptive* (like that of Darwin), and did not take any particular direction (even to "divert" organisms along "aberrant" blind alleys). Confronted with this paradox, Lamarck resolved it by adding to his theory of use and lack of use the idea that all living forms "tended" gradually and inevitably toward ever higher (in other words, more complex) levels of organization. Lamarck never explained the cause of this tendency toward perfection, but later authors attributed it to "vital impulses," and therefore became known as *vitalists*, or *finalists*, because they believed that evolution had directionality.

14

For those who believe that the Story of Life reflects a program unfolding over time, evolution is in some ways comparable to the process of development which leads from the embryo to the adult, following pre-established laws (natural, but which we do not as yet understand very well). Clearly, those who wish to find significance, meaning, or an intention in the Story of Life will always have recourse to mysterious internal forces, still to be discovered or unknowable.

But if it is the case that invisible threads have, from the beginning of time, guided evolution in a linear and orderly manner to result in us, why do we find ourselves among such a diversity of living forms? As we shall see, we are not descended from chimpanzees, but we have a common ancestor with them. Chimpanzees are our brothers, not our parents. Nor are we descended from any organism like a sea urchin. Nevertheless, the phylum of echinoderms, to which sea urchins belong, and that of the chordates, which includes our species, share a very remote ancestor which was neither a sea urchin nor a human being. Living species are not ordered in sequence. What we have here is not a staircase to nowhere, but a tree with many, many branches and with no trunk or central axis. Evolution is not linear, but divergent.

Despite all of this some authors, although aware that, whether looking at the past or the present, evolution does not appear to have moved in one single direction, still express the view that life followed various tendencies, and that ours is the tendency of increasing intelligence. However, they never explain how these tendencies arise: they appear to obey mysterious impulses which have nothing to do with the adaptation of organisms, but rather act of their own accord. The definition of primates still frequently includes the term "tendency toward cerebral expansion," as if a tendency could serve in and of itself to characterize an entire group including fossil and living species, which would thus become a "unit of evolutionary destiny." Naturally, contemporary primates that do not exhibit such cerebral expansion are held simply to represent relics of the past, or "living fossils."

The authors remember a time when groups of primates or hominids which were not in a direct evolutionary line with our species were described as deviant forms; these ideas can still readily be found in serious texts. If such forms became extinct, as a just punishment for their rebellion, they were tarred as abortive or even aberrant forms, described as trials or failed experiments (carried out by whom?), as blind alleys, or in other terms which made it clear that it was not a good idea to diverge from the path marked out by evolution.

But let us leave, for now, the debate between vitalism and materialism, or finalism and Darwinism. Over the chapters of this book we will review what we know today of human evolution through its different stages and circumstances. The last chapter will give us the opportunity to discuss, in the light of past events, the nature of our history. But first, in the next chapter, we will start to get to know ourselves a little better, locating ourselves among the diversity of species of living and fossil primates.

2

We the Primates

*Direct perception of the origins of anything is automatically denied to
our eyes as soon as a sufficient depth of the past is interposed.*
Pierre Teilhard de Chardin, *The Human Zoological Group*

The Ecological Definition and Diversity of Primates

If we observe nature objectively we recognize that, despite its enormous
diversity, there is not an infinite number of forms: in fact, all living species
can be grouped within a limited number of biological types. A lion is
simply a large cat, or conversely, a cat is a small lion. It seems that once
evolution has produced a successful new model for an organism, one
which has been well "tested," as automobile manufacturers would say,
it begins to "manufacture" different variations within the same "range,"
sometimes altering little more than the size. A cat represents the size of
lion appropriate for the hunters of small prey "sector," and thus aspires
to its corresponding "market share," in competition with other small
predators.

Species can thus be grouped in progressively larger categories, such as
Felidae (for all species of the cat type), mammals, vertebrates, and so on.
The larger the category which groups together a given set of species
(felines, mammals, vertebrates, etc.), the further back in time the common
ancestor will be. We belong to the group of primates – the group of
monkeys and apes. It is therefore not correct to say that we are descended
from apes, as if we were no longer apes ourselves. We are still primates just

as much as any of the other approximately 180 living species in the group. However, we evolved not from any current species of ape, but from species that are now extinct, many of which are also ancestors of other modern primates. Having thus situated our species in its natural context, and accepted that we are not an anomaly, let us consider what being a primate consists of, since it is something we cannot avoid.

Primates form a very homogeneous set of species in terms of their ecological requirements. To generalize, they are mammals which live in humid tropical forests (*rainforests*) or monsoon-type subtropical forests, with seasonal rains and dry periods during which some trees lose their leaves. It was in this warm forest environment that we evolved, and all primates therefore show various adaptations to life in the trees.

Of course there are exceptions to this definition. We humans form one such exception although, as we shall see, this has only been the case for a few million years. The baboons developed in the more or less open savanna of Africa, where they accompanied us in our evolution and were probably ecological competitors; at all events, they always seek the protection of groups of rocks or trees at night. A similar species, the gelada (*Theropithecus gelada*), lives on the Ethiopian highlands, far from the trees. The macaques are a set of species found almost exclusively in Asia, some of which live in the cold forests of Japan or in the foothills of the Himalayas, as well as in many other environments. The only African macaque, the Barbary macaque (*Macaca sylvanus*), now lives in the Atlas Mountains of Algeria and Morocco, north of the Sahara. Although the Barbary macaques in Gibraltar were introduced by humans, this species flourished in European ecosystems, settling in latitudes as high as England and Germany before becoming extinct in Europe. Another species which is well adapted to open environments is the patas monkey (*Erythrocebus patas*), which lives in the savanna and only seeks shelter in the trees to sleep or at times of danger. When running, the patas monkey can reach a speed of 55 km per hour, the record for primates.

Currently, apart from ourselves, primates are found in the wild only in Mexico, Central and South America, Africa, and Asia. There are none in Europe or in Australasia.

Primates are fairly varied in terms of type of diet: some species are completely herbivorous while others are omnivorous and also eat small vertebrates and invertebrates like insects; some primates have specialized in eating insects. Baboons and chimpanzees sometimes even hunt and eat other mammals, although their diet is otherwise herbivorous.

Having evolved over a long period while living in the trees, modern primates share a series of unique special features which enable them to grasp and climb trees, and jump from branch to branch. One such adaptation is the first toe (the big toe), which is very large and mobile and can be opposed to the other toes (except in our case). Primates also have flat nails on all the digits of hands and feet, rather than the claws of their ancestors. Some primates have claws, sometimes on all the digits except for the first toe, but it appears that these were originally flat nails which converted back into claws at a later date.

The teeth of mammals are very important in paleontology for two reasons: one is that dentition reflects the type of diet, and the other is that they are the most frequently found fossils – often the only fossils found for certain groups. A first step toward the study of dentition is counting the number of teeth an individual has. All living primates descend from an ancestor who had 36 teeth. However, not all mammals have the same kinds of teeth – they are specialized for different functions. Four groups of teeth are distinguished, from front to back: incisors, canines, premolars, and molars (Figure 2.1). The primate ancestor of all modern primates had, on each side of the mouth and at both top and bottom, two incisors, one canine, three premolars, and three molars. Some primates have modified

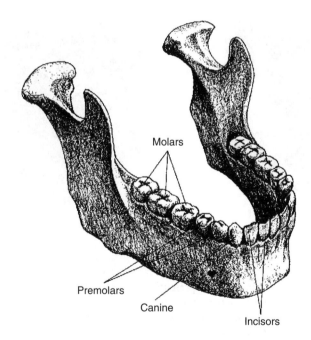

Figure 2.1 Human jaw, showing the four types of teeth

this formula over the course of their evolution, losing some elements, but never increasing the number of teeth in any of the four categories.

The Classification of Primates

Scientists are sometimes thought to study very subtle, obscure character-istics, using microscopes or other advanced instruments to establish the classification of organisms. However, the main division of primates into two large groups known as strepsirrhine and haplorrhine primates (Strep-sirrhinae and Haplorrhinae) is based on nothing more complex than the form of the nose and the upper lip (Figure 2.2).

In strepsirrhine primates, as in other mammals, the external nasal orifices, or *nostrils*, are surrounded by a hairless, moist area of skin known as the *snout*, which continues downward in a lip divided into two parts along its mid-line, through which it is attached to the gums by a membrane. To understand this morphology better, the reader simply

Figure 2.2 Strepsirrhine and haplorrhine primates. *Left*: a ring-tailed lemur (*Lemur catta*), a representative of the Strepsirrhinae; *right*: a pygmy chimpanzee or bonobo (*Pan paniscus*), a representative of the Haplorrhinae. Strepsirrhine primates have hairless, moist skin around the nasal openings, and a divided upper lip attached to the gums along the mid-line. Haplorrhines have the same type of skin around the nasal orifices as on the rest of the face, and the upper lip is undivided and mobile

needs to look at the nearest cat or dog. This arrangement of the nose severely limits the facial expression of emotions. Haplorrhinae, including ourselves, do not have this hairless skin around the nostrils, and the upper lip is undivided and mobile. The continuous upper lip facilitates greater facial expressivity, a well-known characteristic of haplorrhine primates.

The Strepsirrhinae include lemurs, indris, and the aye-aye (lemurs in the broader sense), which evolved and diversified in isolation on the island of Madagascar (off the east coast of Africa). Some species are nocturnal and others diurnal. Unfortunately, the recent arrival of humans has resulted in the degradation of their forest paradise and the disappearance of many species.

The Strepsirrhinae also include the Asian lorises, and the African bush babies and pottos (the Lorisidae in the broader sense), all of which are nocturnal.

The Haplorrhinae are divided into three groups. One of these is the tarsiers (Tarsiiformes), small nocturnal primates found in the Philippines, Borneo, Sumatra, and other Southeast Asian islands, with enormous eyes, very long tails, and very long back legs, well adapted to jumping. The other groups of Haplorrhinae are the catarrhines (Catarrhinae), which include our own species, and the American monkeys, known as platyrrhine primates (Platyrrhinae).

The catarrhines and platyrrhines are usually grouped together in a common category used more or less informally, that of anthropoid apes (technically, Anthropoidea) – from here on we will use the term anthropoid. Both groups are diurnal, with the exception of the South American owl monkey, *Aotus trivirgatus*, which appears to have become nocturnal, although it is descended from diurnal ancestors.

Another feature of anthropoids is the completely frontal position of the eyes, allowing for a wide field of stereoscopic or three-dimensional vision, which requires that the fields of vision of the two eyes overlap. This type of vision makes it possible to calculate the distance to objects, whether tree branches or prey, very precisely. Anthropoids have a large brain, although it appears that platyrrhines and catarrhines developed this (through evolution) separately. The olfactory lobes of their brains are very much reduced in size. As anthropoids, we perceive the world essentially in terms of images, not in odors.

Platyrrhines have the same number of teeth as the first primates, except for the marmosets and tamarins (Callitrichinae), a group which has lost the last molar. The catarrhines, however, have lost a premolar (Figure 2.1), although many of our readers will also find that the last molar (the wisdom teeth) will never emerge. This absence of the third molar in the adult jaw is an expression of the reduction in the dental and masticatory (chewing) apparatus which has arisen in *Homo sapiens*, our species.

Hominoids, Apes of Our Own Branch

Within the catarrhines, our species is classified among the hominoids, while the so-called Old World monkeys form the subgroup Cercopithecidae, which includes the macaques, baboons, mandrills, guenons, colobus monkeys, langurs, and others. In addition to ourselves, the hominoid group also includes a series of primates known as *apes*. These are, in order of closeness of relationship to humans, the two species of chimpanzee (our closest relatives), the gorilla, the orangutan, and the various species of gibbon, the apes furthest from us in evolutionary terms (Figure 2.3). The number of species of gibbon varies from five to nine, according to different authors, as some recognize as distinct species what others consider to be simply geographical variants of a single species. If we have problems classifying living species, imagine the difficulties encountered by the paleontologist working with fossils.

The two species of chimpanzee are the common chimpanzee and the pygmy chimpanzee, or bonobo (Figure 2.2), which is in fact no smaller than the common chimpanzee; the two together form a natural group, or clade, as they are descended from a common ancestor. It should be added, and this is an important detail, that this common ancestor is exclusive to these two species – in other words, no other living species descends from it. The chimpanzees, the gorilla, and ourselves together form another clade: we also share an exclusive common ancestor. Given that chimpanzees and gorillas live in Africa and, as we shall see in this book, we also come from Africa, it seems reasonable to suggest that the still unknown common ancestor lived in Africa.

There has been much discussion as to whether chimpanzees or gorillas are our closest relatives, although genetic studies appear to incline toward the chimpanzee. In fact, the evolutionary line of the gorilla separated from the human line a very short time before that of the chimpanzee, so effectively the three lines separated at virtually the same time. The chimpanzee line later split to produce the common chimpanzee and the pygmy chimpanzee, or bonobo.

As can be seen, the system for classification of species created by Karl von Linné (generally known by the Latin name of Linnaeus, 1707–78) has a hierarchical structure. We are thus first humans, then hominoids, then

Figure 2.3 Types of hominoid. Both sexes are shown. *Top left*: a pair of gibbons (*Hylobates lar*); *top right*: orangutans (*Pongo pygmaeus*); *below (left to right)*: gorillas (*Gorilla gorilla*), common chimpanzees (*Pan troglodytes*), and humans

catarrhines, then anthropoids, then haplorrhines, and finally primates, which in their turn are mammals, then vertebrates, and so on...until we arrive at the largest category, the animal kingdom.

As a group, hominoids share a set of features inherited from our common ancestor. Many of these are related to a particular method of locomotion through the trees which Arthur Keith (1866–1955) termed *brachiation.* This form of locomotion consists of traveling by hanging from branches, with the arms extended, swinging from one arm to the other as the body turns in the air (Figure 2.3). However, there have been many variations within the brachiating model, in addition to those currently found in living species: these represent only a very small part of the diversity of the group in the past, as we shall see. It would perhaps be more correct to say that hominoids show adaptations which allow them to hang from the branches with the trunk erect, rather than traveling over the branches on all fours or jumping from one to the other, as the arboreal (tree-living) primates generally do.

In hominoids the thorax is flattened dorsiventrally (from the chest to the back), rather than being laterally compressed as in the rest of the primates and in quadruped mammals generally (Figure 2.4). In consequence, our shoulder blades (or scapulas) are located in a dorsal position, on the back, rather than laterally, on the sides of the body. The form of the shoulder blades is different, with a longer vertebral edge (the edge running alongside the spine). The form of the humerus (upper arm bone) is also different: the head of this bone (where it articulates with the shoulder blade) is more rounded, and the diaphysis, or shaft of the bone, twists so that the head of the bone faces inward rather than toward the back. The expansion or lateral widening of the thorax also means that the clavicle (collarbone) is longer.

All of these modifications result in a great freedom to move the arms above the level of the shoulders; combined with the capacity to extend the arms completely and the mobility of the wrist, this makes brachiation possible. Since we share these characteristics with other hominoids, we have to acknowledge that our ancestors must have been brachiators.

Also as a consequence of brachiation, the arms are more fully developed than the legs. The ratio of the length of the arms to that of the legs among apes ranges from 147 percent in the siamang (a species of gibbon) to 102 percent in the bonobo. Moreover, the hands lengthen while the length of the first finger (the thumb) decreases, forming a hook from which to hang. This modification of the hand makes it difficult for apes to touch the tip of the index finger and thumb together.

Of course, our ancestors' subsequent adaptation to bipedal walking caused some of the characteristics seen in apes to alter, as we shall see in the chapter on human locomotion (Figure 2.4). In particular, the pelvis and legs have changed markedly, so that the ratio of the length of the arms to that of the legs is only 72 percent, while the big toe is no longer opposable and is aligned with the rest of the toes. Moreover, the thumb has become longer and the rest of the hand shorter, so that we have recovered the ability to manipulate small objects which other hominoids have partly lost.

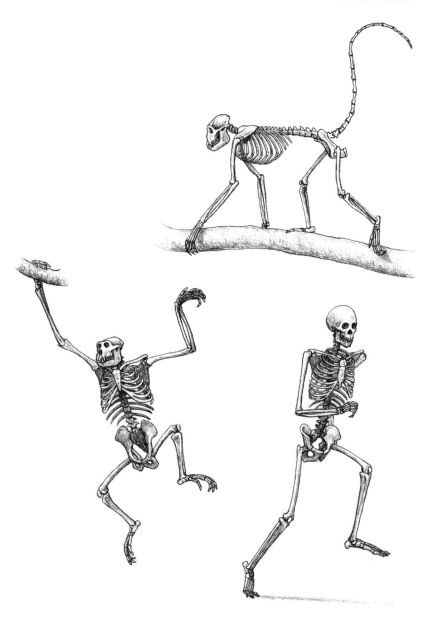

Figure 2.4 Skeletons of primates. *Top*: colobus monkey (*Colobus guereza*); *below left*: common chimpanzee; *below right*: human

In hominoids the lumbar region of the trunk is shorter, with a smaller number of vertebrae (particularly in gorillas and chimpanzees), making us unable to bend this area as much as other primates can. And as we all know, we have no tail, although we are not the only primates without one. Hominoids characteristically hold the trunk erect or vertical, both when climbing and when swinging through the trees or resting.

Gibbons are sometimes bipedal and travel over the branches, rather than hanging from them as they habitually would; in this case they use their long arms to balance. Gibbons rarely come down to the ground, spending most of their lives in the highest level of the forest. Orangutans are effectively four-armed rather than four-legged, using their feet, which are very similar to their hands, to grasp branches or hang from them, and not traveling much through the trees (Figure 2.3).

When they come down to the ground, orangutans, chimpanzees, and gorillas move on all four limbs, but the trunk does not become horizontal as it does in the predominantly quadruped primates: instead it slopes downward from the shoulders to the hips. When moving over the ground, which they do very occasionally, orangutans support themselves on their fists.

In fact, adult gorillas are so heavy that they can barely be considered brachiators (Figure 2.3). The gorillas' and chimpanzees' method of locomotion on the ground is very particular (Figure 2.5): they support

Figure 2.5 Quadruped locomotion of the chimpanzee. *Detail*: position of the bones of the hand

themselves on the soles of their feet and on the dorsal (back) side (not the knuckles, as is often stated) of the intermediate phalanges of the second to fifth fingers (the index to the small finger). Quadruped primates walk on the soles of their feet and the palms of their hands. Gorillas' and chimpanzees' particular mode of travel has also resulted in modifications to the bones of the arms, to give them more stability when supporting the weight of the body on either one.

History of the Primates

The first known fossil thought to be that of a primate (though this is not certain) is a molar from the late Cretaceous, the last period of the Mesozoic or Secondary era, when the dinosaurs were still alive. It is around 65 million years old and was found in Montana. It was named *Purgatorius ceratops*, as it is contemporary with the well-known three-horned dinosaur *Triceratops*.

Purgatorius is allocated to the group Plesiadapiformes, which continues into the following era, the Cenozoic, made up of the Tertiary period together with the Quaternary, or Pleistocene. Plesiadapiformes (Figure 2.6) are the only primate fossils known from the Paleocene, the first epoch of the Tertiary (between 65 and 55 million years ago); during this epoch they split into a number of lines. The last of these became extinct in the subsequent, Eocene epoch (between 55 and 36 million years ago). The fact that Plesiadapiformes have been found in both North America and Europe indicates that the two continents were joined together before they became definitively separated by the Atlantic Ocean. Throughout the Cenozoic, Asia and North America were also joined from time to time, in the region of what is now the Bering Strait. However, Africa was isolated, like a large island, during most of the Cenozoic, although there were sometimes land bridges between Eurasia and Africa which allowed for interchanges of fauna.

There is ongoing debate as to whether the Plesiadapiformes should be considered true primates or not (for example, they do not appear to have had flat nails, nor an opposable big toe). Nevertheless, Plesiadapiformes are the group closest, in evolutionary terms, to living primates, which themselves form a natural group with an exclusive common ancestor. Some authors suggest that primates can be divided into two large categories, Plesiadapiformes, or archaic primates, and others, the euprimates or "true" primates (see Figure 2.7).

Figure 2.6 Plesiadapis. A Paleocene European plesiadapiform. The skull of *Plesiadapis* (*top*) is compared with that of a red squirrel (*below*), showing their similarities – which are due to a similar lifestyle rather than any evolutionary relationship

Euprimates appear in the fossil record in the Eocene (the epoch between 55 and 36 million years ago), and are well represented in Eurasia and North America by two large groups, the Adapiformes and the Omomyidae, although both groups probably also lived in Africa. The Adapiformes show some general similarities with lemurs and lorises, but we cannot say for certain that they are the direct ancestors of these animals (Figure 2.8). Many authors consider the Omomyidae to be haplorrhines, but this classification is also open to doubt. Finally, some middle Eocene fossils from China and Algeria (about 45 million years old) have been identified as anthropoids, though not all authors recognize them as such.

The transition from the Eocene to the subsequent epoch, the Oligocene, is well represented in the El Fayum deposits in Egypt (Figure 2.9). Numerous anthropoid fossils, between 30 and 37 million years old, have been found here. The anthropoids, let us recall, are the group formed by the platyrrhines, or New World monkeys, together with the catarrhines, which in their turn are divided into the Old World monkeys and the hominoids; the latter comprise apes and humans.

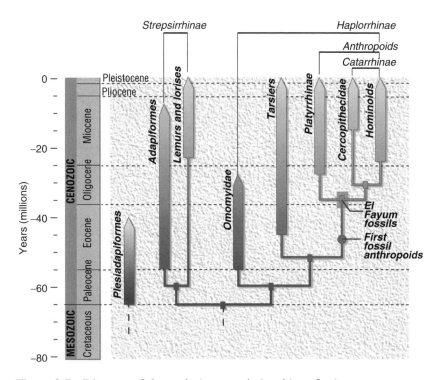

Figure 2.7   Diagram of the evolutionary relationships of primates

How platyrrhine primates arrived in South America is a big question, which has yet to be resolved. At that time South America was an island, separated from North America by the Caribbean Sea (without the Panama isthmus), and from Africa by the Atlantic Ocean. Given that no anthropoid fossils have been found in North America, it seems unlikely that they could have come from there. The Atlantic Ocean had not yet expanded to its present size, and it is possible that at least one pair of primates could have traveled accidentally from Africa, on a natural "raft" formed by trees tangled together, if they had encountered favorable currents and winds. (This is how the ancestors of the lemurs, indris, and aye-aye now found on Madagascar must have arrived on the island: Madagascar has been separated from the African continent for more than 100 million years.) However that may be, the oldest known platyrrhine is *Branisella boliviana*, which is around 27 million years old, and it remains a mystery as to how it arrived in South America. There is another group of mammals that reached the continent in similar, and similarly unknown, circumstances:

Figure 2.8 Notarctus. A North American Eocene adapiform, with hind limbs longer than its front limbs and a capacity for jumping similar to that of the modern lemurs of Madagascar

these are the caviomorph rodents, which include guinea pigs, capybaras, chinchillas, viscachas, and so on.

During the Miocene, the epoch which succeeded the Oligocene and lasted from 24 million to 5 million years ago, the first hominoids – the group to which we and the apes belong – emerged on the African continent. The oldest fossils, found in East Africa, have been assigned to the genus *Proconsul* (Figure 2.10), a type of hominoid which lived before the line which led to the gibbons split off. The name *Proconsul* (pre-Consul) was coined in 1933 in honor of a famous chimpanzee at London Zoo called Consul, in the mistaken idea that the fossil species was the direct ancestor of modern chimpanzees.

After *Proconsul* come *Morotopithecus*, *Afropithecus*, and *Kenyapithecus*. Fossils of these four species dated between 23 and 14 million years ago have been found. *Proconsul* does not appear to have developed the locomotive adaptations characteristic of modern hominoids. However, some details of the lumbar anatomy and of the joint between the shoulder blade and humerus of *Morotopithecus*, 20.6 million years old, suggest that

Figure 2.9 Oligocene fossil anthropoids from El Fayum. *Foreground: Aegyptopithecus and its skull; background: Apidium*

this genus may offer the oldest fossil evidence of a body organization similar to that of living apes.

Around 17 million years ago, Africa (at that time joined to the Arabian peninsula), which had remained quite isolated since the Eocene, drew

Figure 2.10 Proconsul. The first fossil hominoid, which lived in East Africa during the early Miocene

closer to Eurasia; the Indian subcontinent also became connected to Eurasia. This allowed hominoids to spread throughout the Old World and to diversify widely into many genera – each one comprising a number of species – such as *Dryopithecus, Sivapithecus, Lufengpithecus, Ourano-pithecus, Ankarapithecus*, and *Gigantopithecus*. Fossils of these genera have been found in Europe, China, Turkey, India, and Pakistan, and are dated at between 13 and 7 million years old. There seems to be general agreement that there is a relationship between *Sivapithecus* and the orangutans, but the evolutionary position of the other Eurasian homin-oids is the subject of much debate. Louis de Bonis and George Koufos maintain that *Ouranopithecus*, found in Greece (and also known as *Grae-copithecus*), is related to the group comprising ourselves, the chimpanzees, and the gorillas.

In the Can Llobateres deposit, in Catalonia, Spain, Salvador Moyà-Solà and Meike Köhler recently discovered a large part of a skeleton of

Dryopithecus laietanus. With its long arms and short legs, this fossil gives us an idea of the way in which these hominoids, of about 9.5 million years ago, moved around (Figure 2.11). They appear very similar to modern orangutans, which hang from the branches of trees and move slowly through them. A well-known but enigmatic long-armed primate known

Figure 2.11 *Dryopithecus.* Fossil hominoid which lived in Europe during the Miocene

as *Oreopithecus*, which lived during the same period in the swampy forests of central Italy and Sardinia (which at that time formed a single large island), could be close, in evolutionary terms, to *Dryopithecus*. However, in view of the very specific characteristics of its dentition (type and arrangement of teeth), some authors do not even recognize *Oreopithecus* as a hominoid.

The fossil trace of Eurasian hominoids disappears about 7 million years ago, until forms already very close to the modern orangutan and gibbons appear; the one notable exception is *Gigantopithecus*. A species of this genus (*Gigantopithecus blacki*) survived in China and Vietnam until only a few thousand years ago, and therefore must have lived at the same time as humans. We have three jawbones and more than a thousand individual teeth from this species. The first fossils of *Gigantopithecus blacki* to be found were four molars which the paleontologist Ralph von Koenigswald (1902–82) purchased in drugstores in Hong Kong and Canton between 1935 and 1939 (in traditional Chinese medicine, fossils are credited with healing properties). The *Gigantopithecus* jawbones are larger than those of gorillas – particularly one of them, believed to be that of a male. These were probably the largest primates which ever existed.

The front teeth (incisors and canines) were relatively small, while the premolars and molars were large and wide, with a thick layer of enamel. The jawbones are also very robust. In these characteristics they are similar to *Paranthropus*, a type of hominoid with very strong teeth and jaws, which we shall describe later. This similarity has led some people to believe that *Gigantopithecus* belonged to our evolutionary group, but in fact this is a case of adaptive convergence – in other words, a resemblance due to more or less similar chewing activity in two independent lines. The evolutionary affinities of *Gigantopithecus* are more likely to be with *Sivapithecus* and other species in the orangutan line.

In view of their size, we can only assume that these enormous apes lived on the ground, and that they fed on what must have been very abundant plant resources. The strength of the jawbones and the type of dentition indicate that they ate a hard, fibrous type of plant which required much chewing. Some authors believe that this may have been bamboo, eaten by modern pandas.

3

Climate and Evolution

The hay appeareth, and the tender grass sheweth itself, and herbs of the mountains are gathered.

Proverbs 27:25

The Origin of Species

The origin and the disappearance of species are two of the fundamental problems of evolutionary biology. By selecting animals and plants, man has managed to produce many different breeds and varieties but not, so far, any new species. All breeds of a given species can interbreed, producing fertile offspring which can reproduce again. This is not the case with different species of animals: although two different species can be bred together, such as horses and donkeys, their offspring – in this case the mule – is sterile. Obviously this may be a matter of time, given that we have only been domesticating animals and plants on a large scale for 10,000 years.

During his famous voyage of five years and two days around the world in the brig *Beagle*, Charles Darwin began to get an idea of a basic mechanism for the development of new species. If a few individuals remained isolated in a marginal region, they could adapt to the conditions prevailing there and transform into new species. This would explain the diversity of chaffinches found in the Galápagos Islands (Ecuador). Each island had its own distinct species of chaffinch – in some cases more than one – with different specializations (or, as an ecologist would say, occupying different niches); all developed from a single species, which had come from the American continent.

The emergence of new species is not always the result of a seed or a pair of a particular animal species being passively transported to a remote place by winds or currents, as occurred with the Galápagos chaffinches or the platyrrhine monkeys of South America. The area of distribution of a species may be split by a new geographical barrier, giving rise to new species on either side of the barrier. And there are also cases where species are subjected to a change of landscape without having to move or be isolated by changes in topography. This is the case of the hominoids, which lost their environment because of a great, planet-wide ecological change.

Climate Changes over the Last Few Million Years

Climate change is currently a topic of great concern. We fear that human activity may lead to the earth becoming too warm. It seems to us that it is hotter than when we were young. We hear that the desert is advancing and the ice on the mountains and at the poles is melting. In short, the subject of climate change has become fertile territory for the doomsayers of the end of the second millennium. Scientists, however, are obliged to analyze problems and their causes from a wider point of view. In historical geology, widening the perspective means expanding the window of time by several million years (see Figure 3.1).

We have been in a warm epoch for some ten thousand years; this has made possible the current expansion of humanity through the development of agriculture. We should not forget, however, that this is an interval within the cold period of the last million years. Moreover, even within the last ten thousand years the climate has not been absolutely uniform. There have been times when it was much colder than now, and times as hot as or hotter than the present day, but these smaller warm and cold cycles lasted only a few centuries, and were minor in effect.

About 150 years ago we came out of a cold period known as the Little Ice Age, which began in the 15th century and was a major influence on many historical events. During the warm Middle Ages, on the other hand, not only were wine grapes and other sensitive crops often grown in regions of Great Britain that are now unsuited to them, but the

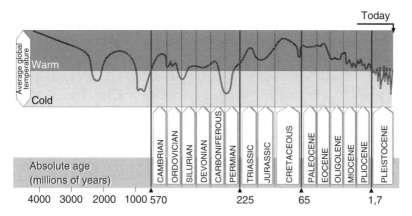

Figure 3.1 Historical variation in global average temperature on earth. There has been a steady decrease in average temperature since the Miocene, with marked fluctuations in the Pleistocene which correspond to the ice ages

Vikings were able to colonize the southern coasts of Greenland for several centuries, growing cereal crops, maintaining farms, and even establishing a permanent episcopal diocese. The name itself, Greenland, gives an indication of what the region was like. However, all traces of the Vikings disappeared from Greenland at the beginning of the 16th century, when the wave of cold became so intense that in London, King Henry VIII was able to cross the frozen Thames River in his carriage.

These small-scale fluctuations are nothing, however, compared with the great climate changes which frame human evolution over the last four or five million years. Over this period, continuing right up until today, we see a general tendency toward a cooling of the planet, combined with an overall decrease in precipitation. These phenomena have not occurred uniformly, but have fluctuated in climatic cycles, progressively more marked toward the present day.

Like the seasonal changes which take place each year, climate changes affect regions in the high latitudes more than equatorial regions, where much of human evolution occurred. In North America and Eurasia the Ice Ages were very pronounced during the last million years, and during this period great sheets of ice extended over a large part of these continents at regular intervals of about 100,000 years.

Paleotemperature Scales

The spectacular traces left by the ice during its advances and retreats in the northern hemisphere have been used to establish a scale of changes in temperature in Europe over time. Nevertheless the Ice Ages, although they were large-scale climatic episodes, were also local phenomena which affected different regions in different ways, even within the small European continent. Moreover, it is difficult to track them further back than one million years.

The scales now used are based on *marine paleotemperature curves,* derived from investigation of the sedimentary deposits accumulated in the seabed. These marine records go back much further and are more continuous than the continental records, and are a more reliable reflection of global changes of temperature.

The most commonly used scale is based on the oxygen found in the minute fossil shells of microorganisms known as foraminifera. These minuscule protozoa live in the sea, and when they die their shells gradually accumulate at the bottom. Dozens of test wells have been drilled in deep water all over the world in order to obtain a microfossil sequence covering a long period of time. In nature, oxygen occurs in two different isotopes,[1] both of which are stable: oxygen-16, light and very abundant, and oxygen-18, heavier and very rare. The ratio of the two oxygen isotopes in seawater, and in the carbon dioxide dissolved in it, depends on the temperature. Because this oxygen passes into the shells of the foraminifera while they are alive, the temperature is recorded in their bodies for all time; when they die this information is gradually deposited at the bottom of the sea.

In order to relate these marine scales, which tell us the temperature of the seawater, with continental climatic cycles, the deposits of dust transported from the land by the wind have also been traced in the seabed, in order to see how these deposits have changed; it is assumed that the drier

---

[1] Isotopes are atoms of a specific element which are chemically distinct from one another. Because the small difference occurs in the nucleus of the atoms (in the number of neutrons, to be precise), it has little effect on the normal properties of the element, except for a minute difference in weight. The fundamental importance of isotopes, and the reason they are so well known, is their nuclear properties: some are stable, like carbon-12, and others are radioactive (i.e., they decompose naturally by emitting radiation), like carbon-14. Some, such as uranium-235, are used to make atomic bombs, while others, like uranium-238 and others, are not.

and more arid the adjacent continental regions, and the less vegetation they have, the more dust will have blown toward the ocean. Fossil pollen grains in the seabed have also been studied, in order to find out what kind of plants covered the surface of the earth.

Marine records tell us what happened over the last few million years, but they do not tell us why. What are the main factors contributing to the changes in the earth's climate?

Factors in Climate Change

The basic factors leading to climate change can be grouped into five main categories: (a) catastrophic events; (b) geodynamic evolution of the planet; (c) behavior of the hydrosphere-atmosphere system; (d) natural fluctuations in the earth's orbit around the sun; (e) the effect of the biosphere, including human activity.

These factors, some of them very intricately enmeshed, produce very different effects. Catastrophic events, which are sudden and unpredictable, such as the impact of giant meteorites or huge volcanic eruptions, lead to marked changes of short duration; only if the change produced is extremely drastic will it affect entire species.

At 10:00 a.m. on August 27, 1883, the largest recorded explosion in history, much larger than any nuclear test, occurred in the volcanic crater of Krakatoa (Indonesia). It was heard 3,500 km away; it destroyed an entire island and spewed forth 21 sq m of rock. It shot enormous quantities of gas and ash into the atmosphere, forming a vertical stream 80 km high. The fine dust particles expelled into the stratosphere traveled around the world several times and produced spectacular sunsets for years, even in Europe. The resulting tsunami ("tidal waves") traveled as far as Hawaii and South America. More than 50,000 people died. It is possible that the global temperature of the earth fell by half a degree, but no long-term change occurred; no species disappeared. Currently there is only one catastrophe theory which is taken seriously. This is the hypothesis of a major meteorite impact, which is suggested as the cause of the extinction of the dinosaurs – although this is still the subject of fierce debate.

Geodynamic evolution includes a wide range of phenomena such as a reduction in the flow of heat from the interior of the earth to the surface, shifts in the geographic and magnetic poles, volcanic activity, and vertical and horizontal movements of the earth's crust. This last phenomenon is of fundamental importance.

Over the last few million years continental drift has resulted in an entire continent, Antarctica, being located exactly over the earth's South Pole. The snow that falls here thus accumulates, forming a layer of ice up to 4 km thick. In the northern, or Boreal, hemisphere the huge continental masses of North America and Eurasia have also moved closer to the North Pole. We are currently in a relatively warm period between two ice ages, but during the ice ages the land at high latitudes was permanently covered by ice, as around eighty percent of Greenland still is; the ice there reaches depths of almost 3 km. At the North Pole there is no land on which snow can accumulate, but as the Arctic Ocean is very enclosed, a permanent, though not very thick, ice cap has formed which floats on the surface of the ocean.

The hydrosphere-atmosphere system is extremely complex. Water's enormous capacity for storing heat means that the oceans act as vast thermostats, moderating terrestrial fluctuations in temperature. In addition, the seas largely control the amount of water vapor and carbon dioxide present in the atmosphere. Precipitation (rainfall and snowfall) depends on the amount of water vapor in the air; in addition, these two gases are the main ones responsible for the so-called "greenhouse effect."

Milankovitch Cycles

The earth's orbit around the sun results in two extremely regular temperature cycles which we know well. The first is the alternation of day (warm period) with night (cool period), caused by the rotation of the earth about an imaginary axis passing through the North and South Poles; as we know, this occurs every 24 hours. The second is the annual succession of seasons, which in the northern hemisphere gives rise to the series spring (temperate), summer (warm), fall (temperate), and winter (cold). This second cycle is due to the inclination of the earth's axis of rotation with respect to the plane of its orbit, an inclination which is currently about 23.5°.

There are four events which mark the beginning and end of the seasons, as we know them in the temperate zones of the earth: two equinoxes (spring, or vernal, and autumnal), and two solstices (summer and winter). At the equinoxes day and night are of exactly equal length. In the northern hemisphere, the summer solstice sees the day with the most hours of daylight in the year, while the winter solstice sees the longest night (see Figure 3.2). On the human scale, these are the only cycles we perceive, and they are too short to leave geological traces.

The hypothesis that large-scale climate changes were due to natural fluctuations in the earth's orbit was first put forward in the 19th century. However, the development of a quantitative theory relating the orbital movements of the earth, the level of solar radiation, and the earth's climate was the impressive work of engineer Milutin Milankovic (1879–1958), who dedicated over thirty years to the study of these phenomena.

If the earth's orbit around the sun was exactly circular (which it "virtually" is), if the sun was exactly in the geometric center of the orbit (which it "virtually" is), and the inclination of the earth's axis of rotation

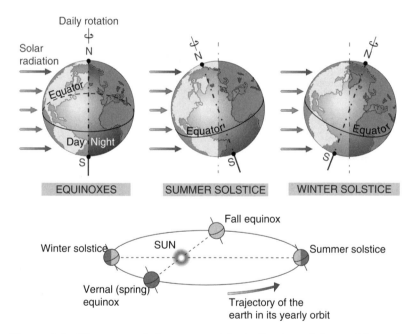

Figure 3.2 The cause of the seasons. The inclination of the earth's north–south axis of rotation in relation to the plane of its orbit around the sun gives rise to the seasons. In the northern hemisphere, the days are long and solar radiation is strong in summer. In winter the days are short, and in addition solar radiation meets the earth at a tangent (note that the opposite occurs in the southern hemisphere). Note that the seasons are not directly related to the earth's greater or lesser distance from the sun

was always 23.5° (as is "virtually" the case), there would be no large-scale climate changes caused by fluctuations in the level of solar radiation: all summers would have been the same for thousands of millions of years. However, all those "virtually"'s added together, along with some others we have not mentioned, result in very gradual changes in the amount of solar radiation reaching the earth each year. According to Milankovic's theory, this is the slow but inexorable motor driving climate change.

In addition to equinoctial precession (Figure 3.3), which results in periods of very hot summers alternating with periods of temperate

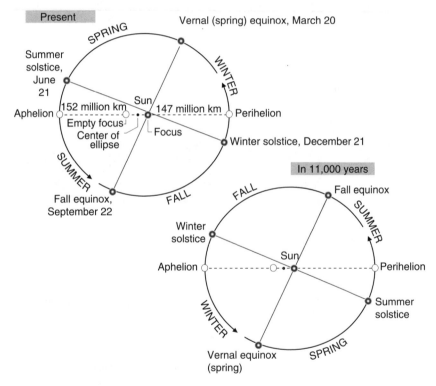

Figure 3.3 Precession of the equinoxes. The position of the equinoxes and the solstices which mark the beginning of the seasons changes slightly each year. Currently, winter in the north begins when the sun is closest to the earth (the perihelion); winters are therefore not particularly cold. In about 11,000 years' time the winter will occur when the sun is farthest away (the aphelion), and northern winters will be harsher. Thus equinoctial precession periodically tempers or heightens the intensity of the seasons

summers approximately every 11,000 years, Milankovic looked at two other, slower fluctuations in his paleoclimatic analysis. The first of these was the change in the inclination of the earth's axis of rotation relative to the plane of its orbit, which is currently around 23.5°, and varies between 21° and 24.5° over a period of about 41,000 years. The second, the slowest of all, with cycles of between 100,000 and 400,000 years, is related to changes in the shape of the orbit (whether it is more rounded or more elongated). All of these orbital variations combine to give rise to complex fluctuations over time in the amount of solar radiation the earth receives.

Since the development of marine paleotemperature scales in the 1970s, particularly the scale based on oxygen isotopes, we have undeniable evidence that climate fluctuates in cycles which coincide with those predicted by Milankovic's theory. All we are now missing is detailed knowledge of the mechanisms which amplify the slight variations in solar radiation levels predicted by the theory, resulting in major changes in climate.

A Climatic Model for Equatorial Africa

At the present time rainfall in subtropical Africa is markedly seasonal, and follows the yearly cycles of the African monsoon. During the northern summer the land in the interior of the continent is heated up, attracting moisture-laden air from the equatorial Atlantic. This results in abundant rainfall in the western and central regions of subtropical Africa. In eastern Africa rainfall is always much lower, because in addition to the reduction we would expect owing to the greater distance which the clouds have to travel, there is a mountain barrier which prevents them from reaching this region. In winter atmospheric circulation is reversed, and cold, dry winds from the northeast blow over the whole of the region. Once again, the effects are more serious in eastern Africa; in the west, local currents continue to bring warm, moist air from the Atlantic.

As a result of these conditions, the vegetation in western subtropical Africa still consists of humid forest today. East Africa, on the other hand, is much more arid. It now has savanna-type ecosystems, with grassland predominating over trees. We shall return to this geographic factor, the cause of such a marked division between the eastern and western regions of subtropical Africa, when we look at what is known as the "East Side Story" hypothesis.

Marine records show that the seasonal summer monsoon already existed 5 million years ago; however, more favorable overall conditions in terms of temperature, humidity, and atmospheric levels of carbon dioxide (CO_2) meant that more or less humid forest covered the whole of subtropical Africa.

Peter deMenocal has constructed a theoretical model of how different factors have affected climate in the low latitudes of Africa over the last few million years. The model explains how the climate of subtropical Africa can be influenced by climate fluctuations in the north – ice ages. Among other elements, it studies the effect of cooling of the north Atlantic on the monsoon in Africa. According to deMenocal, around 2.8 million years ago a phenomenon occurred which changed the climatic history of the northern hemisphere decisively and had a major impact on the ecosystems within which our ancestors were evolving: this was the beginning of large-scale climate fluctuations, with permanent ice over much of the north during the cold epochs. Marine records in subtropical zones of the Atlantic and Indian Oceans, which surround the African continent, show that these climate fluctuations had a marked influence on the climate in equatorial Africa. Since this time the retreat of the forest in this region, giving way to savanna and grassland, has been unstoppable. It seems that this was the situation between 2.8 and 1 million years ago.

Marine records also show that the influence of fluctuations in the north on the climate of this region of Africa has become stronger over the last million years. DeMenocal's theoretical model can explain the increase in aridity in subtropical Africa and its relation to northern ice ages. The main reason for this, it is suggested, is that the cooling of the waters of the north Atlantic led to a series of cooler, drier African summer monsoons which destroyed the delicate tropical forests of East Africa.

Nevertheless, since astronomical factors have always existed, and ice ages have only occurred in particular eras of the earth's history, we have to recognize that such extreme situations only occur when other circumstances are superimposed on orbital fluctuations. One of these, of course, is the poleward movement of the continental landmasses, which encourages the accumulation of ice and hampers the movement of warm tropical waters. The other is changes in the atmosphere, and most particularly, the influence of CO_2.

The Controversial Gas

The natural presence of water vapor (H_2O), carbon dioxide, and other atmospheric gases ensures that the average temperature of the earth's surface is $15°C$, rather than $-15°C$: if it were not for the natural greenhouse effect, the earth's surface would be a layer of ice! For this reason, when we talk of the greenhouse effect as something potentially dangerous, we should really use the term "heightened greenhouse effect" – in other words, the additional overheating which can be caused by emissions resulting from human activity.

CO_2 is the slightly sour, usually harmless gas that gives us the bubbles in our sodas. It makes up only a very small proportion of the atmosphere, barely 0.03 percent, but it is nevertheless vitally important: not only is it the main agent in the beneficial greenhouse effect (when this is not magnified by human activity), it is also the basic source of organic carbon – the carbon of which all living beings are made.

With the help of sunlight, plants convert water and CO_2 into organic matter through photosynthesis. Most plants belong to a group known as the C3 plants, because they fix carbon dioxide through a mechanism which uses a molecule with three carbon atoms. A minority of plants, mostly grasses with tough, fibrous stems, belong to another group known as the C4 plants, since their mechanism for fixing CO_2 uses a molecule with four carbon atoms. A few cultivated crops, such as maize and sugar cane, belong to this second group.

With the right humidity and temperature conditions, and with abundant CO_2, C3 plants develop much better than C4, but with the current level of CO_2 in the atmosphere, which is very low compared with other geological epochs, C3 plants find it difficult to survive in hot, dry environments. Today, C3 plants still predominate in the temperate and cold climates. In the hot climates of equatorial latitudes, C3 plants grow in abundance in the humid forests and also in the so-called rain forests (where an unlimited amount of water is available). The C4 plants, on the other hand, are grasses and reeds typical of the open, dry, sunny environments grazed in Africa by zebra and antelopes such as gazelles, hartebeest, impala, and gnu, and by elephants, hippopotamus, and other herbivores of the great savannas. As these C4 plants have fibrous, mineral-rich stems they cause much wear on the teeth of the mammals which eat them; for this reason their teeth have high crowns, so that they will last a long time.

In addition to their appearance, C3 plants are chemically distinct from C4 plants, since the C4 plants contain a larger proportion of a particular rare isotope of carbon (carbon-13). Thure Cerling and his colleagues have carried out an exhaustive analysis of the quantity of this stable but rare isotope of carbon in the enamel of fossil teeth from grass-eating animals, particularly Equidae (the group which now consists of horses, zebras, and their relatives, the donkeys and wild asses), but also Proboscidea (the elephant group), some extinct South American mammals known as notoungulates, and other groups of large herbivores. The regions studied were Europe, East Africa, Pakistan, North America, and South America, and the period covered was the last 20 million years. The researchers found that 8 or more million years ago levels of this rare isotope were low in all groups, indicating a world dominated by C3 plants. However, 2 million years later the situation began to change in East Africa, Pakistan, equatorial and South America (in Europe and Africa C3 plants have always been dominant).

Cerling and his colleagues concluded that between 8 and 6 million years ago the concentration of CO_2 in the atmosphere began to decrease – a decrease which has continued up to the industrial era. This resulted in the expansion of open ecosystems dominated by C4 plants, and the reduction of the forested areas. Cerling and his colleagues make the interesting observation that the changes in plant cover which affected extensive regions of the earth were accompanied by major changes in fauna, with the expansion of mammals adapted to open environments. These herbivores, with their high dental crowns suitable for grazing grass, replaced those which browsed the trees.

One final note. The reader will no doubt have realized that the emission of CO_2 into the atmosphere which humans cause by burning fossil fuels (coal, oil, etc.) will also have ecological effects in the future, favoring C3 plants. This is just one of the factors, combined with the heightened greenhouse effect, the hole in the ozone layer, acid rain, and others, on which we are exerting an irresponsible influence, well before we have managed to understand the complex mechanism of the earth's climate.

The End of Paradise

In ecological terms, the progressive decrease in the volume of CO_2 in the atmosphere, combined with climatic factors, led to the belt of hot tropical

Figure 3.4 The giant gelada *Theropithecus oswaldi* (right), shown on the same scale as a female hominid, *Paranthropus boisei*

forest which extended over much of the Old World becoming fragmented and declining from the end of the Miocene, and particularly during the Pliocene and Pleistocene. This loss of habitat probably resulted in the disappearance of many species of hominoids, although this may not have been the only cause of the reduction to the very narrow range which exists today. In the Miocene there were various species of hominoids living within the same region, whereas today no more than two species live together. Another major cause of the decline may have been ecological competition with the other Old World monkeys, the Cercopithecidae, which are much more abundant and varied today.

However, this same climate change led to the appearance and spread of more open ecosystems over much of Africa during the late Miocene and Pliocene, with new species of plants and animals. Among these the hominids (our ancestors and our closest relatives) soon appeared, as we shall see in the next chapter. The ancestors of the patas monkey, and of the baboons and the geladas, also took advantage of this change in environment. One form of gelada, *Theropithecus oswaldi*, developed to enormous size during the Pleistocene (weighing up to 100 kg in exceptional cases), and lived in the same zones as humans, who may even have hunted them and contributed to their becoming extinct (Figure 3.4). Numerous remains of this giant monkey, together with an enormous quantity of stone tools, have been found in the Olorgesailie deposit in Kenya (about 800,000 years old).

As we see, we are far from the only primates to have come down from the trees, or to put it another way, to adapt to a world in which trees had been replaced by grass, and forests by grassland.

4

The Origin of Humanity

So for every man who has ever lived, in this universe, there shines a star.
Arthur C. Clarke, *2001: A Space Odyssey*

Molecular Clocks

Studies in molecular biology indicate that our line separated from the chimpanzee line between 4.5 and 7 million years ago – in other words, at approximately the same time that the gradual decrease in atmospheric CO_2 levels was beginning to affect the African ecosystems, as we saw in Chapter 3. This coincidence makes it tempting to speculate that hominids originated as a direct result of ecological change and the spread of open environments, to which they adapted from the beginning. However, as we shall see later, it now seems that the oldest representatives of our group, the first hominids, were as much forest-dwellers as are today's chimpanzees, and that the gradual adaptation to drier, less densely forested environments occurred later.

However this may be, molecular biologists have calculated this interval between 4.5 and 7 million years using their *molecular clocks*. The basis of these molecular clocks is that the genetic difference between species, such as our own and the chimpanzees, ought to be related to the time which has passed since the two lines separated. In other words, genetic divergence increases with time, as does the morphological difference between two lineages which diverge from one another to follow different evolutionary paths.

49

But this assertion that genetic difference is related to the length of time the lines have been separated is only valid if appropriate genes are chosen for analysis. The genes suitable for use as molecular clocks are those known as "neutral," those which confer neither advantage nor disadvantage, and on which natural selection therefore does not act. In neutral genes, the naturally occurring spontaneous mutations accumulate at a constant rate, without being either eliminated or favored, like snow falling steadily.

On the other hand, the "non-neutral" genes that selection does target may be modified at different, varying rates depending on the intensity of the pressure for selection exerted upon them. In other words, if a particular gene (the correct technical term is an *allele*) is very beneficial for the individual that carries it, it is sure to spread rapidly throughout the population. If, however, it confers a disadvantage, its frequency of occurrence in the population will fall rapidly because it has natural selection, a powerful enemy, against it. Furthermore, what is beneficial today may not be so tomorrow, or may not be so in another species – so these non-neutral genes are of no use for measuring time in evolution. To take an example from everyday life, the power and capacity of personal computers, which are subjected to market pressure for selection, increase very rapidly, and not at a constant rate. This is a clock which gains time.

But in order to calculate the rhythm of change of neutral genes, what is known as the *rate of mutation*, we have to return once again to fossils, measuring the genetic difference between two species for which we know, from fossils, how long their lines have been separated. For example, in order to work out how long it is since the human and chimpanzee lines separated, we can use the human/orangutan pair. Measuring the genetic distance between the two is the easy part, although since not every gene can be used the calculations vary depending on which ones are chosen. Establishing when the orangutan line separated is a very different matter. Sometimes the figure of 13 million years is used: this corresponds to the first fossils attributed to the species *Sivapithecus*, which themselves are believed to mark the beginning of the evolution of orangutans.

In other words, in order for the "molecular clock" to work, many things are needed: genes which natural selection does not "see" but which are known to us, constant rates of mutation, and a good paleontological reference point – too many things – but nevertheless, this interval of between 4.5 and 7 million years for the human/chimpanzee separation is acceptable to paleontologists, as we shall see shortly.

The First Fossil Hominids

Before continuing we should pause for a moment to clarify a question of terminology, in order to avoid confusion. Some paleoanthropologists use the term "hominid" in a very broad sense to refer to humans, chimpanzees, gorillas, and the fossil relatives of the entire group. We prefer to use the term hominid in its more traditional sense, to refer only to modern humans and the fossils of our own evolutionary line, i.e. the species which emerged after the separation of the chimpanzee line. Other authors define hominids as bipedal primates. However, although it is true that all the species with erect posture come under our definition of hominid, as we shall see, we do not yet know for certain whether the first hominids already walked on two feet. All bipeds are hominids, but it may be that not all hominids were bipedal.

A fragment of a mandible (lower jawbone) with a molar found at Lothagam (Kenya), to which has been ascribed a geological age of more than 5.6 million years, may be that of a hominid, although the small size of the preserved fragment makes it difficult to be sure. There are other fossils which similarly offer little information, including a jawbone fragment from Tabarin and a fragment of the proximal (upper) humerus from Chemeron (both in Kenya), dated at about 4.5 million years old.

Aside from these isolated, doubtful remains, the oldest set of hominid fossils yet found was located in 1992 by a team led by Tim White, Gen Suwa, and Berhane Asfaw in the middle reaches of the Awash River in the Afar region of Ethiopia. These Middle Awash fossils have been published only in part, although White and his colleagues have already created a new genus and species for them: *Ardipithecus ramidus* (the words *ardi* and *ramid* come from the Afar language and mean, respectively, "ground" and "root," while *pithekos* means "ape" in Greek). The information that is available on these fossils indicates that they are very primitive forms of hominid, around 4.4 million years old. In fact, they show features so primitive, particularly in their dentition, that it seems likely that they cannot be very far from the division between the chimpanzee and human lines. An age of between 4.5 and 7 million years therefore seems acceptable for the time being, and if we were pinned down we would go for a date closer to 4.5 than to 7 million years. In any case, it is likely that we will have a definitive answer before very long.

It also appears that *Ardipithecus ramidus* lived in a forest environment. This conclusion is drawn first from the type of mammals – forest-dwellers – with which the hominid fossils appear. Monkeys of the colobus type and

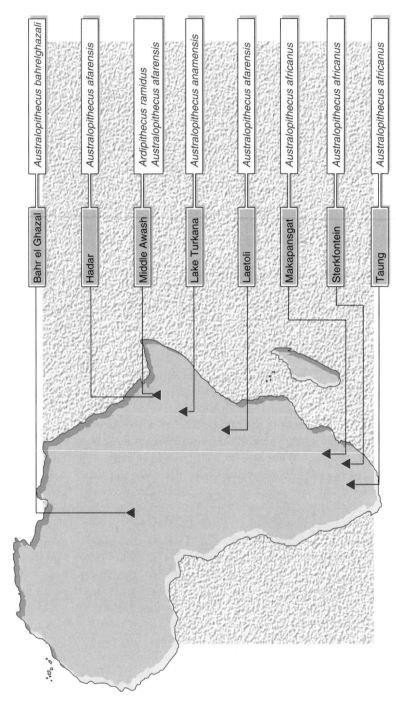

Figure 4.1 Location of the main deposits containing fossils of *Australopithecus* and *Ardipithecus*

tragelaphine antelopes (the group of the kudus, sitatungas, and others with spiral horns) are especially abundant. Secondly, the outer enamel layer on the teeth of *Ardipithecus ramidus* is thin, as in chimpanzees, which eat fruit, leaves, tender stalks, shoots, and other soft plant products. However, the teeth of fossil hominids dating from after *Ardipithecus ramidus* have a thick layer of enamel, protecting them from the wear caused by a diet incorporating tough plant elements such as roots, tubers, grains, nuts, and so on. So it appears that the first ancestors of man, the first hominids, were primates who lived in the forest and had a diet very similar to that of modern chimpanzees. But we shall leave the detailed analysis of hominid diet for a later chapter.

On the basis of some aspects of the base of the skull, in fairly fragmented remains, White and his colleagues have suggested that these first hominids were bipedal and walked as we do. However, this remains to be demonstrated on the basis of hip and leg bones, some of which are known to have been found in the most recent excavations.

Change of Habitat

A Kenyan team led by Meave Leakey (wife of the famous fossil-hunter Richard Leakey, of whom more later), found hominid fossils about 4 million years old (between 3.9 and 4.2 million years old, to be precise) in Kanapoi and Allia Bay, either side of Lake Turkana in Kenya. In 1995, on the basis of these, they identified a species they named *Australopithecus anamensis* (the term *anam* means "lake" in the Turkana language, so the name of this species could be translated "australopithecine of the lake"). These are again very primitive hominids, to judge from a maxilla (upper jawbone) and a mandible (lower jawbone) which have been found, but they have thick enamel on the molars. The association of fossils in which they occur suggests an open forest environment, or more or less wooded savanna with streams. There are also colobus monkeys and spiral-horned antelopes, but these are associated with other species more common to open environments, such as jerboas, a type of mouse from the arid steppes.

The hominid remains found include a tibia (shinbone), which its discoverers believe indicates that these primates were bipedal. We are witnessing for the first time, about 4 million years ago, the appearance of hominids who have begun to change their way of life, their environment, and their diet significantly, and who also move in a completely new way.

Although hominids are not the only primates living in the savannas and grasslands, the erect posture is a completely new innovation.

The next million years (roughly speaking) correspond to another East African species, known as *Australopithecus afarensis* (which can be translated as "australopithecine of the Afar region"). Most of the fossils of this species have been found in the Hadar area, on the lower reaches of the Awash River (in the Afar region, Ethiopia), and in Laetoli (Tanzania).

The Laetoli fossils, which include a mandible classified as the type specimen (holotype) of the species (known as L.H. 4), are dated at 3.5 million years old; the Hadar fossils are between 3 and 3.4 million years old. Donald Johanson's team, working in Hadar since 1972, has discovered numerous remains, so that *Australopithecus afarensis* has a reasonably complete fossil record. This includes the skull of a male (A.L. 444-2), discovered by Yoel Rak in 1992 (Figure 4.2), and a large part of the skeleton of a female (A.L. 288-1), known worldwide as Lucy, as she was named by Johanson when he found her in 1974.

An assemblage of fossils which includes the most complete known mandible of this species was found in the Maka deposit, 3.4 million years old, on the middle reaches of the Awash River. Some teeth a little over 4 million years old found in Fejej, southern Ethiopia, and a fragment of frontal bone from Belohdelie in the Middle Awash, approximately 3.9 million years old, have also been attributed to *Australopithecus afarensis*. However, the assignation of these fossils to the species is subject to review and they could, given their chronology, belong to *Australopithecus anamensis*.

The associations of vertebrates in the Hadar deposits suggest changes in the paleoecological conditions over the 400,000 years of geological history recorded there. *Australopithecus afarensis* appears to have lived both in a fairly dry forest environment and in a cool savanna with corridors of forest along the river valleys – in other words, neither a humid forest nor an arid steppe, but rather an intermediate habitat.

One of the fundamental problems of paleontology is that of grouping fossils into species, since unfortunately the remains do not appear with labels in the deposits. It is the paleontologist's job to find their place in evolution.

Living species which are related to one another are often morphologically very similar, or may differ only in external features such as color, fur, behavior, and other characteristics which, although they may be very striking, are not reflected in the skeleton, the part that becomes fossilized. Ian Tattersall has pointed out that many modern species of primates would not be recognized as different species if we only considered their

Figure 4.2 Male *Australopithecus afarensis*

skeleton; we may thus be seriously underestimating the number of fossil species, since we might group together, as one species, two species which had very different external features in life, although they had an identical or very similar skeleton.

Conversely, great variation can occur within one species when there are major differences between males and females. This differentiation between the sexes is known as "sexual dimorphism," and can affect size, shape, or both. The paleontologist could thus make the mistake of assigning to two different species fossils which simply represent different sexes of the same species.

In short, these problems are often the subject of major debates among specialists. *Australopithecus afarensis* is no exception, and when the species was identified in 1978 by Donald Johanson, Tim White, and Yves Coppens, some researchers did not accept that all the fossils from Hadar and Laetoli should be grouped together in a single species – albeit a very variable one with major sexual dimorphism. Many paleoanthropologists saw two species in this set of fossils, rather than just one, although they did not agree on which fossils should be allocated to which species. Earlier we mentioned the skull A.L. 444-2, which is large and could be a male of the same species as Lucy, a small individual who may have been female. The alternative is that these represent two different species. We chose this example deliberately because, as often happens in paleontology, two different parts of the skeleton are being compared: the skeleton of the body (or postcranial skeleton) of A.L. 444-2 is not preserved, while only a small part of Lucy's skull has been recovered. In addition, Lucy is 3.2 million years old, while A.L. 444-2 is around 200,000 years younger.

Some experts have found significant differences in the postcranial skeleton, and on this basis they identify a completely bipedal species at Hadar and Laetoli, one they suggest is directly related to us and our species, rather than being an ancestor which combined the capacity to walk on two legs with the ability to climb trees. Finally, some people have even suggested that the females were lighter and were climbers, while the males were heavier and bipedal!

There is one very convincing argument for the unity of *Australopithecus afarensis*, based on the fossils found at Hadar site A.L. 333. Numerous hominid remains from 3.2 million years ago have been found here, with virtually no bones of other animals. These hominid fossils represent at least thirteen individuals of different ages, who might have died together in a natural disaster such as a flood. It is very possible that they formed part of the same group and therefore the same hominid species. In fact, this handful of fossils is known colloquially as the "first family."

If the sample from deposit A.L. 333 had included only large individuals, or only small ones, or individuals with a particular morphology, then we might conclude that fossils of different species had been artificially included in *Australopithecus afarensis.* Conversely, if the sample included all the sizes and morphotypes found at Hadar and Laetoli, we could be sure that *Australopithecus afarensis* is a real species and not a hodgepodge of odds and ends. In fact, there is wide variation within the A.L. 333 sample – as wide as has been suggested by the researchers who identified *Australopithecus afarensis.*

East Side Story

We have discussed what is known about when hominids first appeared; now we need to examine where this happened: where was our first home? As we have seen, the fossils of the first hominids were found in East Africa. More specifically, these fossils were found along the Great Rift Valley, an enormous, widening fracture in the earth's crust which runs from Mozambique, through Malawi, the Great Lakes region, the Afar region of Ethiopia, and the Red Sea to arrive at the Dead Sea, between Israel and Jordan.

The geographical distribution of the first hominid fossils suggests that our group originated in East Africa – this is the theory that Yves Coppens has called "East Side Story." According to this hypothesis, throughout the Miocene there was a great belt of tropical forest extending from the Gulf of Guinea across to the Indian Ocean. The tectonic process which formed the great, continent-spanning fracture of the Great Rift Valley must have brought with it changes in relief, raising great mountain barriers and high plains. From the end of the Miocene these separated the eastern ecosystems, with their increasingly open environments inhabited by hominids, from the western ecosystems, which had humid forest environments and were populated by the ancestors of chimpanzees and gorillas.

This hypothesis becomes even more attractive when we consider that it would mean that our emergence was not a unique phenomenon. We are simply part of a community of animal and plant species – a biota, to use the biogeographical term – characteristic of an entire region, and linked to the geological and climatic history of that region. Now, although this hypothesis appears very reasonable, it starts from the basis that the first hominids originated in East Africa. It is in this region that the oldest fossils known up to this day have been found. However, at the end of 1995 Michel Brunet and his colleagues published the discovery north of N'Djamena, in Chad, of the front portion of an australopithecine mandible and an isolated premolar, which are dated to between 3 and 3.5 million years old on the basis of the accompanying fauna. The authors suggested, in a subsequent study, that this was a species different from *Australopithecus afarensis*, which they named *Australopithecus bahrelgazali* (the term *bahrelgazali* refers to the region of Chad in which the fossils were found – Bahr el ghazal, which in Arabic means "river of the gazelles"). This discovery suggests that, early on in their history, hominids spread far to the west of their East African birthplace – if this is in fact where they originated.

The African deposits we have discussed thus far, which are lacustrine or fluvial sediments (from lakes or rivers), usually contain aquatic species such as turtles, crocodiles, fish, and hippopotamus. This does not help us to determine in what kind of ecosystem the hominids lived: other animals, not aquatic or amphibian, are more useful to us. This is no simple matter, for a sedimentary basin may contain the remains of animals that lived and died in very different environments, and have been transported there by streams and rivers. Thus, everything accumulates at the bottom of the basin, creating many problems for paleontologists, who try to resolve them principally on the basis of common sense. The application of common sense to the study of the formation of such deposits constitutes a paleontological discipline in itself, a very important one, known as *taphonomy*. Using taphonomy we can establish, for example, whether a bone has been transported over a long distance or whether the animal died close to where it is found. Fortunately, the study of the adaptations found in fossil species, or functional paleomorphology, also contributes to determining the place of these species in the ecosystem (their niche), and what their environment was like.

We shall return to the question of the environment in a later chapter when we look at the diet of the first hominids. For now let us take a moment to consider the fossils associated with *Australopithecus bahrelgazali*. Michel Brunet imagines that these australopithecines lived in a variety of environments, including forest corridors (characterized by the presence of bushpigs), wooded savanna with elephants, and grasslands where rhinoceros grazed.

The main differences between the Chad fossil and *Australopithecus afarensis* are in the interior face of the symphysis, or anterior (front) part of the jawbone, which has a fairly flat and vertical surface, without the strong transverse reinforcements or tori (ridges – from the Latin *torus*) characteristic of other australopithecines. In this respect *Australopithecus bahrelgazali* is similar to our genus, the genus *Homo*. However, all the premolars have three roots – a primitive trait. To further complicate matters, the jawbone of *Australopithecus bahrelgazali* has wide premolars. This expansion of the premolars is, as we shall see below, typical of some later hominids known as *Paranthropus*.

This combination of features may not constitute sufficient grounds for considering the Chad fossil as a different species, but if it does, we are presented with a new situation. For the first time in our evolutionary history, two species of hominids would coexist, albeit in different regions. While human evolution has traditionally been explained as a linear succession of species, we shall see that the evolutionary tree in general – and our case is no exception – has many branches, even when in some cases only one branch (like ours) ultimately reaches the present day.

Dating Fossils

The reader has perhaps already wondered how we know the age of these fossils of our very distant ancestors. The paleontologist Yves Coppens, who gives many lectures, is often asked this question, to which he recommends the reply: "Trust us, we know the age of fossils, we have methods for determining it, we are professionals." At the risk of boring the reader, we shall disregard his advice and attempt a brief discussion of this fundamental question.

Because of the intensity of the internal forces which come into play, fracturing of the earth's crust is often accompanied by volcanic activity, which may cause ash to be thrown up into the air during the course of eruptions. Winds and water transport the ash, which is finally deposited in beds intercalated between the layers of sediment which contain the fossils. These layers of volcanic ash, or tuffs, are very useful in correlating and dating formations. Even two successive eruptions from the same volcano, with very little time between them, have distinct characteristics, known as their "chemical fingerprints." Using chemical analysis, we can compare two layers of volcanic tuff and find out whether they are the same, even if they do not run continuously in the field because the rocks are fractured into blocks, as is typical of the geology of rift valleys.

During the 1980s a method of dating these tuffs was invented, based on melting a single small crystal of a mineral of the potassium feldspar group using a laser beam. When the crystal melts, the laser releases a given quantity of the gas argon, which is measured with an instrument known as a mass spectrometer. The isotope argon-40 derives from the decay of a radioactive isotope of potassium (potassium-40) contained in the mineral. When the mineral was formed it contained only potassium and no argon. Since radioactive decay occurs at a known, constant rate, the final ratio of radioactive potassium to argon gives us a very reliable age for the tuff. A variant of this technique, known as argon-39/argon-40 dating, is what is now used.

Another method used to date volcanic rocks is the *fission-track* method. The decay (fission) of radioactive uranium (uranium-238) produces marks (tracks) in the crystals of certain minerals, such as zircon. The density of these tracks depends on the quantity of uranium in the mineral and the time elapsed since the volcanic eruption in which the mineral formed.

The fossils lying between successive tuffs in sedimentary sequences can be dated with a precision unimaginable a few years ago. The Middle Awash is an extremely fortunate case, where the majority of the fossils of *Ardipithecus ramidus* come from sediments enclosed, like a geological sandwich, between two volcanic tuffs, one below and one above. Both tuffs are approximately the same age, 4.4 million years old, and this is also the age of the fossils found between them (unfortunately, this is not the general rule, and often there are hundreds of millions of years between tuffs lying above and below fossils).

The argon-39/argon-40 and fission-track methods are used with volcanic materials which, although frequent in East Africa, are not found in by any means all of the formations containing hominids. Other radiometric methods use other isotopes, like carbon-14 or the uranium series. The carbon-14 method (the first to be developed) can be used only for organic matter (matter of animal or plant origin), and is very reliable. Unfortunately, even with the latest improvements, it cannot date anything older than 50,000 years.

Caves frequently contain speleothems (stalactites and stalagmites), which form through continuous precipitation of calcium carbonate dissolved in water. If we are lucky and the carbonate crystals are sufficiently pure, speleothems can be dated using the uranium series method, up to a maximum of about 350,000 years.

But in many cases we do not even have datable speleothems in the formations. Two related techniques, known as *electron spin resonance* (ESR) and *thermoluminescence* (TL), have been developed for these cases. ESR is usually used on the tooth enamel of various mammals, while TL dating is applied to charred flint tools, and a type of sediment which has been exposed to sunlight. The basis of both techniques is that minerals such as flint or quartz, like teeth and bones, act as natural Geiger counters, accumulating the radiation received over time.

Another method used in the measurement of geological time (geo-chronology) is *paleomagnetism*. The earth acts as a magnet, creating a geomagnetic field around it, with two poles (north and south). This field orients the compass needle, indicating the position of the magnetic north pole, which is today close to the geographic North Pole. But it also orients ferrous clay minerals, like tiny compasses, provided that they are deposited slowly in a calm environment, such as an undisturbed pool or lake.

Over extremely long periods, the magnetic poles exchange position (so that the magnetic north pole is located close to the geographic South Pole). The location of the poles at any given moment is recorded in the clay minerals making up the layers of sediment. These changes in the earth's magnetic polarity have been dated, enabling us to establish a scale showing the sequence of alternating periods of one or other type of polarity. Each of these long periods of time is called a *chron*. A *subchron* is a shorter unit of time, with a polarity opposite to that of the chron within which it occurs; even shorter periods are known as *excursions*. Paleomagnetism cannot give us an absolute date for a formation, but it can help with dating when combined with other methods.

In addition to these methods of dating, the fossils of the animals associated with the hominids also help to establish their relative age, since the evolution of species means that animals, including hominids, change over time. Thus biochronological scales can be created, which can be calibrated with the absolute dating obtained by physical methods.

The Taung Child

November 28, 1924 was a great day in the history of paleoanthropology. On that day Raymond Dart (1893–1988), a young professor of anatomy at the University of Witwatersrand in Johannesburg, South Africa, re-ceived a package sent from the Taung quarry containing a child's skull, in which Dart recognized a very remote ancestor of ours. He created a new species and a new genus for this creature: *Australopithecus africanus* (we have waited until now to explain the meaning of the term *australopithecus*, which is made up of the terms *pithecus*, or "ape," and *austral*, meaning "south"). The formation where the skull was found was destroyed and did not yield any further hominid fossils, but two other quarries,

Sterkfontein and Makapansgat, proved very "productive" of fossils of *Australopithecus africanus.*

Thanks to the Sterkfontein Formation, we have an extensive record of *Australopithecus africanus* (Figure 4.3), including a very complete and emblematic skull (found in 1947), referenced as Sts 5, and popularly known as Mrs Ples (Ples being an abbreviation of *Plesianthropus,* the genus to which this specimen was at first ascribed, although it was later realized that it was identical with *Australopithecus*). Another important skull is Sts 71, also found in 1947; in 1989 a very complete skull (Stw 505), apparently male, was recovered, but this has not yet been published in detail. Many pieces of postcranial skeleton have also been recovered, the best known being skeletons Sts 14 (discovered in 1947) and the recently discovered Stw 431. In fact these South African limestone quarries are caves filled with fossils and a much-hardened sediment, forming a very hard breccia (coarse-grained rock) which renders the extraction of fossils extremely difficult.

In the region of South Africa where all of these fossils were found there are no layers of volcanic ash to enable us to date the fossils by radiometric methods. We have to rely on the evolution of the animals accompanying the hominids. By this method, it has been established that *Australopithecus africanus* lived in South Africa between 3 and 2 million years ago. The Makapansgat Formation appears to be the oldest, and its fossils would appear to be chronologically very close to the youngest fossils of *Australopithecus afarensis.* The Sterkfontein hominids might be around 2.5 million years old, with the Taung Child being the youngest representative of the species. The environment in which the Sterkfontein australopithecines lived is interpreted as being forested, though not humid – it seems more likely that it was dry woodland or scrub with open spaces. In other words, a mosaic of ecosystems.

Distinguishing Marks

The distinguishing features of our species are a large brain, a unique capacity to create various kinds of tools using widely varying materials, articulated language, a long childhood, implying a long learning period, and a bipedal mode of locomotion (as well as a very specific sexuality, which we will consider later). The characteristics of large brain, slow development, and capacity to use or adapt natural objects to make tools are also found in our closest relatives, the chimpanzees, gorillas, and

Figure 4.3   Female *Australopithecus africanus*

orangutans – naturally at a much lower level of development, but comparatively greater than that of other animals. These features, plus the capacity for language, can be grouped under the label of something we understand intuitively, but which is impossible to define or measure, and which we call *intelligence*. Locomotion is another question, and since Darwin, science has been asking whether increased intelligence preceded the erect posture, or vice versa, or whether the two evolved at the same time. Which is the same as asking: What was the initial impulse in our evolutionary history – in other words, what made us human?

Stanley Kubrick's film *2001: A Space Odyssey*, based on the book by Arthur C. Clarke, and hailed as a masterpiece even when it was released in 1968, offered an answer which was very much in line with what was thought in some scientific circles at the time. The film depicts a group of apes which can be recognized as hominoids. They are not bipedal, and they are shown against a background of savanna – presumably in Africa – very arid, almost desert. These creatures shelter at night in caves to protect themselves against leopards, and fight over a water pool with a rival group of apes. In other words, they do not yet show any of our distinguishing features. Suddenly they find themselves in front of a monolith of extraterrestrial origin, which they touch. Then comes the spark which initiates human evolution: they have an idea. This idea is to use an animal bone as a tool. For what purpose? In order to kill, in an orgy of blood, first an animal, and then their enemies in the rival group. In other words, our ancestors discovered technology and became both carnivorous and killers of their fellows.

The idea that the first hominids were hunters, or to put it more crudely, "murdering apes," was developed by Dart during the 1950s. Dart believed that the australopithecines were hunters and cannibals and, most importantly, that we have inherited the heavy burden of those violent instincts, at the same time as improving on their weapons. For the australopithecines, according to Dart, did not have dressed stone tools; instead they used weapons made from bones, teeth, and animal horns, an industry that Dart called "osteo-donto-keratic," in reference to these three types of material. According to this theory, it was the hunting instinct and the taste for flesh that led the early hominids to leave the trees and make their first weapons, refining their intelligence and also favoring the adoption of the erect posture, which was certainly more appropriate for a warrior than the quadruped trot. It is worth noting, in passing, that in the past, reflections on human evolution frequently suggested that intelligence developed as hominids were confronted with new challenges – first the savanna and later the colder climates of Europe – but stagnated among those who opted for the "comfort" of the forest and its abundant fruits, or the warm African continent.

Although Dart's interpretation of our origins was not universally accepted, he was not alone. In a famous book entitled *Man-apes or Ape-men: The Story of Discoveries in Africa*, published in 1967, Sir Wilfrid Le Gros Clark (1895–1971) opined that the australopithecines were too poorly defended, with their small canine teeth, to survive without weapons, whether of stone, bone, horn, or teeth. According to Le Gros Clark, these hominids were hunters and carrion-eaters cast into a hostile environment; however, walking on two feet left their hands free to manipulate tools, and it was this that was the stimulus for the development of intelligence. Le Gros Clark was a great authority on human evolution, and it was he who, following a trip to South Africa in 1947 to see the original fossils, helped to change the opinion of the majority of the scientific community, allowing the australopithecines to be recognized as primitive members of our line.

In short, the hunter hypothesis suggests a somewhat bloody beginning to human evolution, though it is nevertheless a beginning. But was this how things began? Have we been "murdering apes" and makers of tools from the start, perhaps even before we became bipedal?

In order to answer this question, we shall address the question of australopithecine locomotion in the next chapter. Later we shall look at other forms of hominids, *Paranthropus*, and the first humans, together with the first stone tools, and then we will discuss evolutionary changes in intelligence, diet, growth, and sociability.

5

The Bipedal Chimpanzee

Moreover, if the individuals I am talking about, moved by the need to grow higher so as to see all at once far and wide, were forced to hold themselves upright and acquired from that a constant habit from one generation to the next, there is no doubt once again that their feet would have insensibly taken on a shape appropriate for holding them in an upright position.

Jean Baptiste de Lamarck, *Zoological Philosophy*

The Great Step

The human being is not the only mammal capable of walking on its hind limbs. We have already noted that anthropoid apes are in the habit of maintaining their trunk vertical as they move through the trees hanging by their arms, or simply when they are sitting. But holding the trunk erect is only half the battle in achieving erect posture and walking on two legs. The other half involves aligning the legs with the trunk – in other words, extending the entire body. The great apes sometimes walk on two legs, but although they hold their trunk almost vertical, they keep the hip and knee joints flexed, just as when they walk on four legs. Only we humans are able to take stable steps without large movements of the trunk, and long strides when we walk, extending our legs far behind the hip; other mammals take only small, wobbly steps, with large shifts of the trunk.

Part of the reason for this major difference lies in the pelvis. When we are standing still, more or less steady, the body is stable and the pelvis is horizontal. However, at the moment when we move a leg forward to take a step, the weight of the body tends to cause the pelvis to lean over toward the unsupported side of the body, putting the walker at risk of falling. But this does not happen, because humans have muscles known as abductors, which stabilize the pelvis and prevent it collapsing too far toward the unsupported side (Figure 5.1).

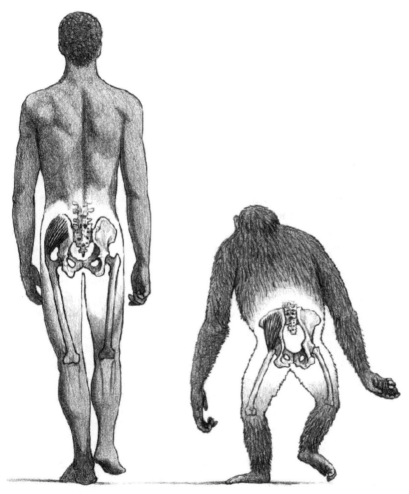

Figure 5.1 Position of the gluteus medius muscle in humans and chimpanzees. In humans the muscle fibers are oriented laterally, and the muscle therefore acts as an abductor, balancing the pelvis when the opposite foot moves forward. In chimpanzees the fibers are oriented toward the back, so that the muscle acts to extend the joint between the pelvis and femur

By contrast chimpanzees, for example, have no mechanism for stabilizing the pelvis, and in order to avoid falling sideways they have to shift the whole trunk a long way toward the supporting side, so that their walk becomes extremely swaying and impractical. On two legs, chimpanzees walk in a way similar to humans who have suffered paralysis of their abductor muscles. We therefore have to ask whether nonhuman mammals do not in fact have pelvic abductor muscles, and when and how these muscles appeared in human evolution.

Although only bones become fossilized, and flesh never does, paleontologists are able to study the function of muscles of which nothing remains. This is the province of paleobiomechanics, a branch of functional paleomorphology which applies the principles of mechanics to the body's levers in order to reconstruct the movements which beings of the past were able to make. The absence of muscles in fossils is not an irredeemable loss, because in fact there are no muscles which specialize exclusively in abduction, nor in adduction (the opposite movement), nor in flexion or extension (the opposite of flexion), nor in rotation.

Muscles simply contract when they receive a neural impulse. That is all. The effect produced by this contraction on the body's mechanism depends exclusively on the line of action of the muscle. And to determine the line of action of a muscle, all we need to know is its two points of contact with the skeleton, points known as the origin and the insertion of the muscle. Of course, the actual movement may result from the activity of several muscles which produce the same action (synergistic muscles), or different and even opposite actions (antagonistic muscles). But as a whole, an organism's capacity for movement can be established in all its complexity if we know the lines of action of all the muscles.

In our species the two pelvic abductor muscles are the gluteus minimus and, most especially, the gluteus medius. The gluteus maximus, which forms most of the muscular mass of the buttocks, works to extend the joint between the pelvis and the femur (naturally there is one each side, as in all other paired muscles). Its action is to align the trunk with the legs. In a standing person, the gluteus maximus straightens the trunk. We can also say that in humans the pelvic extensor muscles are also pelvic stabilizers, but not transverse stabilizers like the abductors: rather they are anteroposterior stabilizers (from front to back). In humans the gluteus maximus is not involved in normal walking on a flat surface, but it comes into action when we run, jump, or go up a hill or up stairs (Figure 5.2).

In a quadruped the extensors of the pelvis–femur joint perform the important task of extending the two hind limbs alternately and pushing the body forward on four feet. Sprinters use the same extension when they

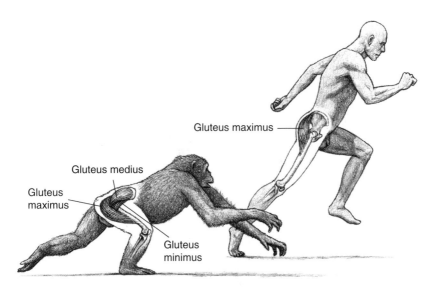

Figure 5.2 Extensor muscles in chimpanzees and humans. In chimpanzees the three gluteal muscles act as extensors of the pelvis–femur joint. In humans only the gluteus maximus does this

push themselves out of the starting blocks, where they position themselves in a quadruped posture to begin with, with the hip joint highly flexed, before moving immediately into a bipedal posture.

In apes, as in other mammals, the function of the gluteus medius and gluteus minimus is different from our own, because the muscles have a different line of action. In chimpanzees and gorillas the three gluteal muscles (maximus, medius, and minimus) act as hip extensors, never as abductors; this means that habitual bipedal walking is not possible, but it favors quadrupedal locomotion.

We have already noted that the function of a muscle is determined by its line of action. Why is the line of action of the gluteus medius and minimus different in humans? The answer, to put it simply, is the orientation of the bony region in which both muscles originate, the iliac crest of the pelvis.

The pelvis is composed of the two coxal bones, one on either side, and the sacrum at the back, which in its turn is formed by the sacral vertebrae (the sacrum is thus part of the spinal column and continues into the coccyx). The coxal bone is actually the result of the fusion of three bones during adolescence – the ilium, which forms the upper part of the coxal bone, the ischium, which forms the lower posterior part, and the pubis, which represents the lower anterior part. These three bones meet in the acetabulum, where the coxal bone articulates with the head of the femur (see Figure 10.1).

Of all parts of the postcranial skeleton, i.e., the skeleton excluding the skull and jawbone, the pelvis is probably that which most distinguishes humans from apes. This is obviously the result of our particular mode of walking, since the pelvis of apes is not substantially different from that of other quadruped mammals. Let us look at the main innovations in the revolutionary architecture of the human pelvis, and consider their biomechanical significance.

In quadrupeds the weight of the trunk is transmitted through the four limbs, but in humans, because we are bipedal, it is transmitted through the spinal column down to the sacrum, and from there through the coxal bones to the heads of the two femurs, continuing down to the feet. In walking, during the phases when we are supported on only one leg almost all of the body's weight is transmitted through one coxal bone (on the supporting side). In order to reduce the stress on the bar of bone running from the sacro-iliac joint to the coxo-femoral joint, the two joints have become much closer than they are in chimpanzees and other quadrupeds. This biomechanical improvement of the pelvis has, as we shall see later, an undesirable side effect: it greatly complicates childbirth because it reduces the size of the bony channel through which the full-term fetus has to pass in order to be born.

The three gluteal muscles originate in the iliac crest and insert in the femur. The iliac crest forms the larger part of the iliac bone or ilium, and in apes consists of a high, narrow plate of bone. In humans it is proportionally shorter, because, as we have already noted, the joints between the coxal bone and the spine and femur are closer together, but most importantly, it is relatively much wider than in other apes.

The other major difference in the human iliac crest is its orientation (see Figure 10.1). In quadrupeds the surface on which the gluteal muscles originate faces directly backward. This orientation means that contraction of the gluteal muscles can only produce extension of the hip joint, because its line of action is posterior: in other words, if the legs are stable they draw the pelvis backward, while if the pelvis is fixed it draws the legs backward.

In humans, however, the orientation of the iliac crest has changed radically, so that the surface where the gluteus medius and minimus originate faces toward one side rather than backward. The biomechanical result of this is that when the hip pulls sideways, the contraction of these two muscles produces the effect of abduction, making bipedal walking possible, as the trunk can be balanced at each stride. The gluteus maximus, on the other hand, which originates in the hindmost part of the ilium and in the sacrum, continues to act as an extensor in humans, since its line of action is posterior rather than lateral.

One of the biggest questions of evolutionary biology is how the great anatomical transformations which create organisms radically different from their predecessors come about. A bipedal primate represents a revolutionary change, not simply a slight variation on other types of hominoids. We have to rule out the idea that the entire skeleton altered drastically all at once, but it is not easy to imagine how an organism can move gradually from walking on four legs to walking on two. One interesting hypothesis is that the initial modification which made it possible for primates to begin walking bipedally was a change in the orientation of the iliac crest. Merely altering in this orientation, causing the iliac crest to face more to the side, would generate some capacity for abduction, one of the basic features of bipedalism. If walking on two legs increased the possibilities for surviving and reproducing, other modifications would subsequently continue to be selected until the entire skeleton was affected.

We will not exhaust the reader with any more biomechanical explanations, although this is a fascinating discipline which allows the paleontologist to play God for a moment, and say to a fossil skeleton: "Get up and walk!" Just one final detail on the pelvis: in humans the capacity for extension of the hip, which is very useful for climbing trees and for walking on four legs on the ground, is also reduced in the hamstring muscles, which originate in the lower part of the pelvis (Figure 5.3).

Let us turn now to the fossils. So far we have been engaging in an exercise of comparative anatomy, the best starting point for paleontological analysis. We have a limited sample of fossil pelvises from australopithecines. The most complete is that of Lucy, which has retained the sacrum and all of the left coxal bone, and those of two Sterkfontein skeletons, Sts 14 and Stw 431 (the latter is fairly fragmented and has not yet been analyzed). Although they show some distinctive characteristics, their morphology is entirely human and does not resemble that of apes in any way. The ilium is short and wide (favoring abduction), and the ischium is short (reduced capacity for extension). Furthermore, the sacrum in these pelvises is proportionately wide, not narrow as in apes. Various authors offer different interpretations of the exact orientation of the iliac crest, which has had to be reconstructed in Sts 14 and Lucy because both had been distorted by being crushed in the deposits. Some authors suggest that the iliac crest in these fossils faces less to the side and more toward the back than in modern humans, but all researchers

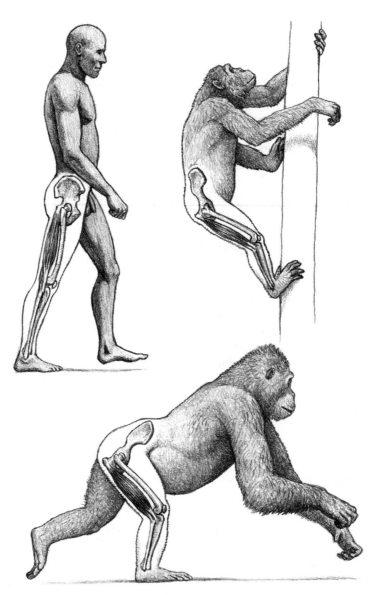

Figure 5.3 Hamstring muscles in humans, chimpanzees, and gorillas. These muscles, consisting of the biceps femoris, semimebranosus, and semitendinosus, make up the muscular mass of the posterior face of the thigh; they are known collectively as the hamstrings, and they run from the lowermost part of the ischium, or ischial tuberosity (the bones on which we sit), to the tibia and fibula. In all cases their action is to extend the joint between the pelvis and femur

recognize that australopithecines had the capacity for abduction of the pelvis which makes it possible to walk on two legs.

There are other skeletal indications that the australopithecines were bipedal. In a human standing stably, the diaphysis (the shaft or main axis) of the femur runs at an angle from the hip to the knee; the knees are very close together (Figure 5.1). If the knees and the feet, which are located vertically beneath the knees, were far apart the center of gravity would have to shift far to the supporting side of the body on each step, and walking would be less efficient and would use more energy. This modern morphology of the femur is also found in australopithecines. However in chimpanzees, for example, the femurs are not inclined downward and inward (quadrupeds do not bring their knees together below their belly, as they have no problems balancing when they walk); this, combined with the lack of abductors, means that when walking on two legs they have to tip the trunk markedly over the supported side of the body with each step, to ensure that the center of gravity does not project beyond the supporting foot. Thus their walking is very inefficient and energy-intensive, as it requires large movements of the body in order to move forward a very small distance.

Another classic feature indicating that australopithecines were bipedal is the position of the *foramen magnum*, the opening at the base of the skull through which the spinal cord runs. In humans the *foramen magnum* is oriented downward because the spine is positioned vertically when we walk. This does not mean that the spine runs in a straight line: in fact in humans (as perhaps in australopithecines too) the spine is markedly curved, with a cervical and a lumbar lordosis (curves toward the front in the regions of the neck and lower back, respectively), and a thoracic kyphosis (curve toward the back in the region of the ribs) (Figure 5.4).

In apes the *foramen magnum* is further behind the base of the skull and is oriented more toward the back, because in quadrupedal walking the head is positioned on the end of a diagonal spine (with no cervical or lumbar curves). The nuchal plane, i.e., the part of the occipital bone in which the muscles of the back of the neck which hold the head in position originate, is also wider and oriented more toward the back in apes than in humans. *Australopithecus afarensis* appears to be at an intermediate point between these features, leading some authors to reconstruct the neck of this species as more inclined than ours, but more vertical than that of chimpanzees.

However, not all the experts agree on how the australopithecines walked on two legs. Some, among which we number ourselves, believe that their way of walking was not substantially different from our own,

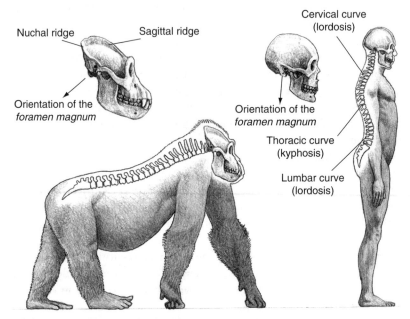

Figure 5.4 Curves of the spine and orientation of the *foramen magnum* in gorillas and humans

while others believe that it was less "perfect." The discussion might have continued ad infinitum had not the team directed by Mary Leakey (1913–96) made a completely unexpected discovery in 1978 and 1979: several meters of tracks formed by footprints left by three hominids who walked through an area of what is now Tanzania about 3.5 million years ago.

The Laetoli Footprints

The Laetoli Beds are located not far from the famous Serengeti National Park. During one of its eruptions a nearby volcano, Sadiman, threw ash out into the air, and rain converted this to mud in which the tracks of many animals were recorded and fossilized. Among these animals were the hominids we have mentioned. There are two parallel tracks, but that on the right (looking in the direction of walking) appears to be that of two individuals, one of them walking in the tracks of the other. The individual on the left is very small, perhaps a female or a child.

The characteristics of these footprints are incredibly modern. They show no sign of insecure or "imperfect" bipedalism; rather they indicate, even in the smallest details, a way of walking identical to our own. The foot of a chimpanzee, gorilla, or orangutan is very different from a human foot. In fact it is more like our hand – flat, with a big toe which is shorter than the others and can separate from them laterally, and extends away from the other toes on each step (Figure 5.5).

In the Laetoli footprints we can study the mode of walking of the hominids which produced them. If you are fortunate enough to have a beach nearby, you can compare these fossil footprints with your own, and appreciate their extraordinary similarity. In each step taken by the Laetoli hominids the front foot was supported first on the heel, leaving a deep impression in the soft ground. Part of the body weight was then transferred through the arch or instep. After this, the foot flexed over the toes, giving a final impulse to lift the foot from the ground and extend the leg forward. As in modern humans, the big toe was fundamental to this final phase (being the last to leave the ground), and it therefore pointed forward like the other toes and was aligned with them (Figure 5.5).

Fossils of *Australopithecus afarensis* of the same age as the footprints have been found in the Laetoli Beds, and we can therefore only suppose

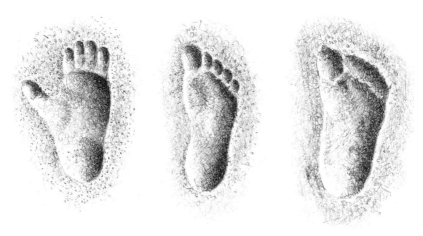

Figure 5.5   Footprints of a chimpanzee (*left*), a modern human (*center*), and a Laetoli hominid (*right*)

that the footprints were produced by three individuals of this species of hominid.

However, the researchers who, on the basis of analysis of fossil bones, had concluded that individuals such as Lucy could not have had a bipedal locomotion so similar to our own (including Brigitte Senut, Randall Susman, Jack Stern, William Jungers, Russell Tuttle, and Peter Schmid), remain unconvinced by the evidence of the Laetoli footprints. Their explanation is that 3.5 million years ago there were two radically different types of hominids in East Africa. On the one hand there were the hominids like Lucy, who were bipedal but had a mode of walking that was not as fully developed as that of humans, and spent much of their time in the trees, where they felt more comfortable; these hominids were close relations, but not our ancestors. Then there were some other, enigmatic, hominids, our direct ancestors, who walked on two feet in the modern way, and it was they who created the Laetoli footprints.

We have already made clear our position on this question. In our opinion, and that of many other authors, the fossils from Laetoli in Tanzania and from Hadar in Ethiopia represent a single species, *Australopithecus afarensis*, which varied widely in size but had a single mode of locomotion. Until new fossils indicating the contrary appear, these australopithecines are the only candidates for having made the Laetoli footprints. In other words, Lucy walked like us.

The Mystery of Mysteries of Human Evolution

Having discussed the origin of one of our principal distinguishing features, the erect posture, there remains one basic and inevitable question: what is it for? The traditional answer was that walking on two legs was an adaptation to the savanna, in order to see over tall grasses or something along those lines. However, there is a conceptual error here. Animals adapt not to environments, such as savanna, forest, or sea, but to ecological niches, to the roles that species play in ecosystems, which may be very varied. To put it another way, there are many species in the modern savanna and none of them is bipedal except for ours. So what we need to ask is what kind of ecological niche the first hominids to become bipedal were occupying. Moreover, it now seems that australopithecines were forest-dwellers rather than inhabiting open environments.

Before returning to this question it is worth pausing for a moment to rehabilitate the erect posture, which is traditionally denigrated and considered inefficient, an evolutionary botch job whose enormous draw-backs had to be compensated by some major advantage. In general it was thought that the advantage was that the hands were freed from walking, allowing for the manufacture of tools and the development of the brain. However, as we shall see, all of these things came long after we became bipedal.

Bipedalism involves an extensive reorganization of the skeleton, and this was achieved with remarkable perfection from the engineering point of view. We can see our body is as if it were built from Meccano: it is made up of a number of articulated segments whose individual centers of gravity are located in the same plane as the axes of the main articulations between the segments (hips, shoulders, elbows, wrists, knees, ankles, etc.). This is at the same time the plane which contains the center of gravity of the entire body. This means that our standing posture is very stable, and requires virtually no effort to maintain. Only the center of gravity of the head is slightly forward of its joint with the first cervical vertebra. This disadvantage has to be compensated by the muscles of the back of the neck, which hold the head erect. However, the reduction of the facial skeleton over the course of human evolution has resulted in a considerable improvement in this prob-lem, shifting the head's center of gravity backward.

One way of assessing the biomechanical efficiency of a way of moving is to follow the trajectory of the body's center of gravity. In our case, the center of gravity is located in front of the second sacral vertebra, more or less at the level of the navel. If we watch a person walking from the side, we will see slight rising and falling movements of the head, corresponding to rises and falls in the center of gravity. Seen from the front, the walker's head inclines slightly toward the supported side at each step.

The straighter the trajectory of the center of gravity, the more econom-ical, in terms of energy consumption, is the walk. If the center of gravity travels on a very winding path, with marked rises and falls and large lateral shifts, the walk will be inefficient and will waste energy. This is what happens when chimpanzees walk on two legs. However, in humans the center of gravity travels on an almost straight line in walking.

So in physical terms our way of walking is no less successful than that of quadrupeds, although of course it is not as fast over short distances. On the other hand we have great stamina, better than that of many quadru-peds, when we need to travel long distances or over long periods of time, whether running or walking. And in fact chimpanzees' and gorillas' way of walking, supporting themselves on the middle phalanges of the hands, cannot be considered a marvel of adaptation to life on land; rather it seems to be a compromise solution to the problem of needing to move both through the trees and over the ground. It appears that gorillas later largely abandoned life in the trees when they became very heavy.

Peter Wheeler has identified a further advantage in the shift of the body to the vertical, related to body temperature. An individual standing on two feet receives less solar radiation, especially when the sun is at its highest point, than a quadruped. Moreover, when the body leaves the ground it moves away from a focus of heat and benefits from cooling breezes. Combining this advantage with that we noted above, we might conclude that bipedal locomotion is perhaps the best solution for a hominid who has to travel long distances while exposed to the sun. The first bipedal hominids did not live in the savanna, but they might nevertheless have had to travel between patches of vegetation separated by open spaces.

One indisputable benefit of bipedalism is the ability to use the hands and arms for carrying, whether food or perhaps children. In this, quadrupeds are at a clear disadvantage to us. We shall return to this capacity for carrying in the chapter on the social biology of the first hominids, because one proposed explanation for the origin of bipedalism is partly based on this factor. Without jumping ahead, we can already anticipate that such an explanation will have many problems, in terms of what we intuit about the social biology of the first hominids. Thus we can still today consider the significance of our acquisition of our special way of traveling as one of, if not the biggest problem confronting human paleontology.

Portrait of the Entire Body of an Australopithecine

Lucy was a very small individual. She was about 105 cm tall and weighed no more than 30 kg. But she was not exceptional among her species. Even the first fossils discovered by Donald Johanson in 1973, consisting of a knee joint (the lower end of the femur and the upper end of the tibia), correspond to an individual of the same size as Lucy; some of the hominids from site 333 are of similar size (as we noted above, this assemblage of fossils was christened the First Family). Other fossils are larger and correspond to individuals around 135 cm tall, weighing about 45 kg. It is thought that the small individuals, like Lucy, were females, and the larger ones males. The male and female averages may have been somewhat closer in *Australopithecus africanus*, but in any case australopithecines were small hominids, similar in size to chimpanzees.

In addition to being small, the proportions of Lucy's limbs were different from ours, including those of modern populations of small stature such as the Pygmies. The most distinctive feature of Lucy's postcranial skeleton is the shortness of the legs (Figure 5.6). The ratio of the length of the humerus to that of the femur (which expresses the relation of arm length to leg length) is 85 percent, markedly higher than that of humans, which averages 71 percent, and markedly lower than that of orangutans (129 percent) and gorillas (118 percent), but not so different from common chimpanzees (102 percent) and bonobos (98 percent). Chimpanzees have relatively longer forearms than we do: the ratio between the ulna (one of the bones of the forearm) and the humerus is around 80 percent in humans, and around 95 percent in chimpanzees (even higher in orangutans and gibbons). In Lucy the ratio is estimated at 92.5 percent.

These proportions suggest that the australopithecines still maintained some of their ancestors' aptitude for climbing trees. Moreover, the phalanges of the fingers and particularly of the toes of *Australopithecus afarensis* are curved, resembling those of chimpanzees more than any other hominid; this curvature of the phalanges relates to the capacity for grasping branches and moving through the trees. All of these factors, combined with the more or less enclosed forest habitat in which they are thought to have lived, leads us to believe that in addition to being able to walk on two legs, australopithecines still climbed trees to find food, to escape predators, or to sleep. Chimpanzees use branches and leaves to construct a kind of personal nest in the tops of trees where they can spend the night. Gorillas also do this, although on the ground, because of their weight. It would not be surprising if the australopithecines still retained this habit of sleeping in the trees (see Figure 11.2).

Despite the extraordinary amount of information which Lucy and other fossils give us, there are still many features of the australopithecine body as a whole which remain unclear. We have already noted that in the great apes the lumbar section of the spine is proportionally shorter. This means that the last ribs are very close to the pelvis. The great apes therefore lack our characteristic narrowing of the waist. Furthermore, in chimpanzees, gorillas, and orangutans the rib cage widens toward the base, taking the form of a cone (see Figure 2.4).

In humans the lumbar section has become a little longer, and as the ilium has also become much shorter we have a capacity to rotate the thorax (ribcage) at this level. This is very useful to us in bipedal walking, since as we bring the hip forward on the side of the front foot, the shoulders turn in the opposite direction. Moreover, our ribcage is barrel-shaped (Figure 2.4).

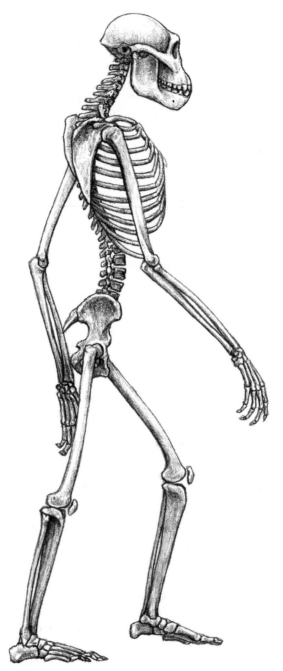

Figure 5.6   Side view of a male skeleton of *Australopithecus afarensis*

In australopithecines the marked shortening of the ilium seems to have clearly separated it from the lowest ribs (Figures 5.7–5.10). Furthermore, there were more lumbar vertebrae (five or six) than in the great apes (which have three or four). All of this meant a greater capacity for turning at the waist. However, the shape of the rib cage might have been more like that of the great apes.

To complete the full portrait of the australopithecine body, we still need the head. If we want to picture one of these small bipedal hominids, it is best to start by giving it the head of a chimpanzee or a female gorilla. From here we have to shade in the details. The skull is divided into two parts – the facial skull, the skeleton of the face, and the cerebral skull or neurocranium, in which the brain is encased. The australopithecines' brain was little larger than that of chimpanzees, and therefore the neurocranium would not be much bigger. There are, however, some differences. We have already noted the position of the *foramen magnum*, which is related to posture and distinguishes the australopithecines from all the apes, although the neck was somewhat inclined and the head further forward in the first hominids (Figure 5.6).

In male orangutans and gorillas, the temporal muscles (which are involved in chewing, and of which we will speak more later) are so highly developed that they need a greater bony surface than is provided by the walls of the neurocranium, and thus a bony ridge forms over the whole of the top of the cranium, along the middle or sagittal plane (Figure 5.4). At the back this sagittal ridge merges with another, transverse ridge which forms the upper limit of the nuchal plane and results from similarly highly developed muscles in the back of the neck. These ridges and the muscles originating in them give the heads of male gorillas, the old silverbacks, a pointed appearance, which the reader has perhaps noted. Sagittal and nuchal ridges are not frequently found in male chimpanzees, but they are found, although in more attenuated form, in some examples of *Australopithecus afarensis* which have been judged to be male, such as A.L. 444-2 (Figure 4.2). It is not clear whether male *Australopithecus africanus* had similar sagittal ridges. Another feature distinguishing australopithecines from apes is the brow ridge. This is a bony rim above the eye socket, which in chimpanzees and gorillas is clearly marked and separated from the rest of the frontal bone by a furrow, to the point where it can clearly be seen in a living animal (see Figure 2.2). Australopithecines do not have a brow ridge as such: the bony rim is not separated from the frontal bone by a furrow (compare Figures 4.2 and 4.3 with Figure 7.4).

The facial skeleton of australopithecines appears little different from that of chimpanzees and gorillas. The face of australopithecines projected

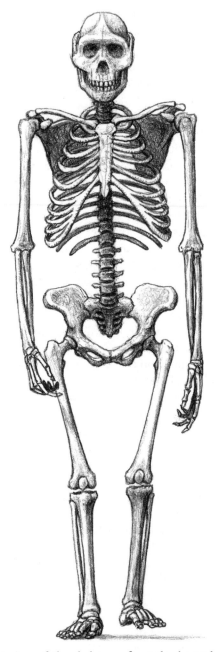

Figure 5.7 **Front view of the skeleton of a male** *Australopithecus afarensis.*
The rib cage is reconstructed as funnel-shaped, as in apes

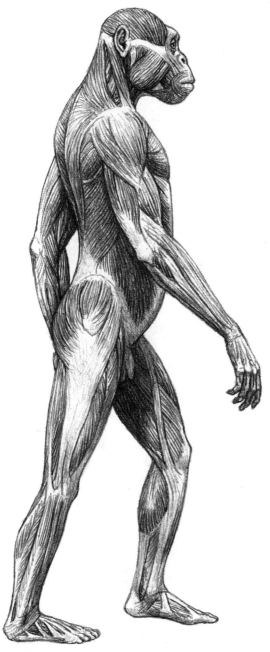

Figure 5.8   Reconstruction of the musculature of a male *Australopithecus afarensis*

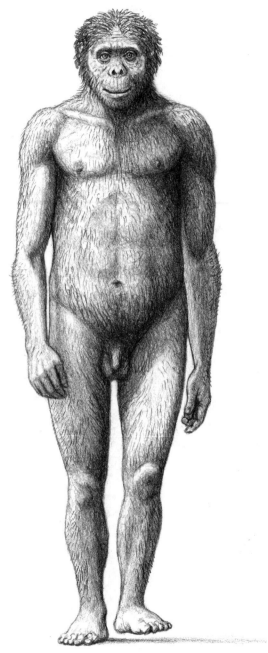

Figure 5.9   Impression of a male *Australopithecus afarensis*

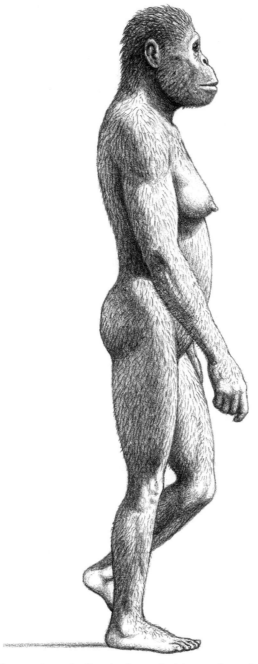

Figure 5.10   Impression of a female *Australopithecus afarensis*

well forward, showing a high degree of prognathism, to use the technical term: the nose did not stand out from the face because the nasal bones did not form a bony bridge.

There are of course many other distinguishing features which are of interest to the specialist, but basically, the head of an australopithecine is thought to have been generally similar to that of an African great ape of the gorilla or chimpanzee type, but with its own particular features – among which the small canine teeth are perhaps the most notable.

6

Paranthropus – Hominids of the Open Plains

Gert was found, and drew from his trouser pocket four of the most wonderful teeth ever seen in the world's history.
Robert Broom, *The South African Fossil Ape-Men*

The Emergence and Distribution of *Paranthropus*

As we have already noted in a previous chapter, marked climatic fluctuations, associated with the expansion of the ice sheets in the northern hemisphere during the cold eras, began to occur about 2.8 million years ago. This change in climate had important consequences for the flora and fauna of the East African equatorial regions, where grassland spread to replace other, more wooded environments. The small mammals and bovids (especially the antelopes) found in faunas of this era provide a record of how forms typical of humid forest environments were replaced by other species more adapted to open, arid environments; these are still found in the savanna today.

On the basis of these data, Elisabeth Vrba has suggested that this climatic and ecological crisis could have had a decisive influence on the evolutionary history of hominids, contributing to the disappearance of *Australopithecus africanus* (still linked to forest environments), and favoring the selection of new forms adapted to life in more open environments – the first representatives of the genera *Paranthropus* and *Homo* (humans).

The term *Paranthropus* literally means "beside man," and was coined by Robert Broom (1866–1951) in 1938 to name a series of hominid

87

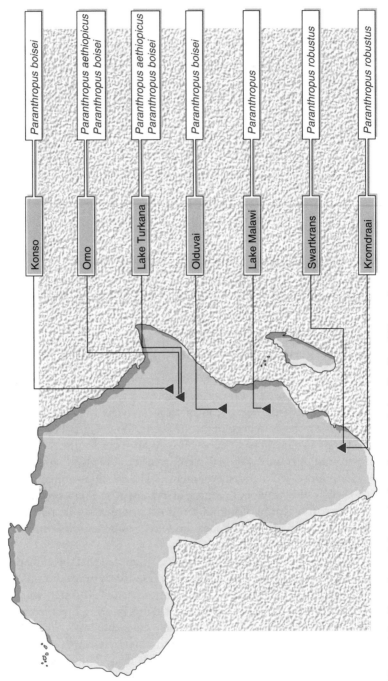

Figure 6.1 Location of the main deposits containing fossils of *Paranthropus*

fossils he had found in the South African Kromdraai deposit. The term is very appropriate, since *Homo* and *Paranthropus* emerged during the same era and lived alongside one another for about 1.5 million years, inhabiting a vast region extending from present-day Ethiopia to the southern tip of Africa.

The majority of paleontologists believe that there were three species within the genus *Paranthropus*. Two of these are exclusively East African: *Paranthropus aethiopicus* (*aethiopicus* refers to Ethiopia, the country where the first remains of this species were found) and *Paranthropus boisei* (the term *boisei* is in honor of an English sponsor named Boise); the third has been found only in deposits in the south of the continent, and is known as *Paranthropus robustus* (*robustus* meaning "robust" in Latin).

Although the first remains attributed to the genus *Paranthropus* – specifically to *Paranthropus robustus* – appeared in South Africa, the oldest known fossils were found thousands of kilometers away. In 1967, Camille Arambourg (1885–1969) and Yves Coppens published details of a hominid jawbone (Omo 18-1967-18) found in one of the deposits of the lower reaches of the Omo River (close to where it flows into Lake Turkana), in Ethiopia; the age of these fossils was established as 2.6 million years. This discovery passed almost unnoticed until, in 1986, Richard Leakey and his collaborators published the find of an extraordinary skull (WT 17000) belonging to the same species as the jawbone found by Arambourg and Coppens two decades before, in one of the deposits on the eastern shore of Lake Turkana. This fossil is one of the amazing "missing links" which paleontologists are occasionally lucky enough to come upon, since it documents an evolutionary stage intermediate between *Australopithecus afarensis* and *Paranthropus boisei*.

Dated at around 2.5 million years old, WT 17000 is the most representative fossil of the species *Paranthropus aethiopicus* (Figure 6.2). Many of the features typical of the anatomy of *Paranthropus* are not yet present in this species, while others are only in the incipient stages of development. The most recent fossils attributed to *Paranthropus aethiopicus* are a series of remains of skulls and mandibles found in the river Omo deposits, and dated at around 2.3 million years old.

The first fossil of *Paranthropus boisei* was found in 1959, in Olduvai Gorge, Tanzania, by Mary Leakey and Louis Leakey (1903–72), Richard Leakey's parents. It consists of an almost complete skull which its finders dubbed "Dear Boy," since it belonged to a young specimen. The age of this fossil (known as OH5) is around 1.8 million years, and the Leakeys attributed it to a new species and genus, *Zinjanthropus boisei*, although it came to be included in the *Paranthropus* genus soon afterward.

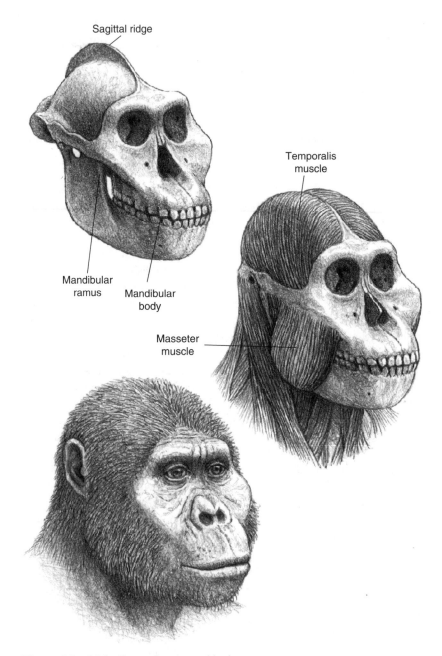

Figure 6.2   Male *Paranthropus aethiopicus*

Paranthropus boisei is now a relatively well-known species (Figure 6.3), whose oldest fossils consist of a fragment of jawbone and some teeth about 2.3 million years old, found in the deposits of the Omo River valley. The youngest known fossils of this species (two milk teeth), which are about one million years old, also come from these deposits.

The most extensive collection of fossils of *Paranthropus boisei* has come from deposits on the eastern and western shores of Lake Turkana, lying within the Koobi Fora stratigraphic formation, and has been built up by Richard Leakey's team from the late 1970s onwards. In addition to large numbers of teeth and jawbones, some skulls have been found, and among these that known as ER 406, ascribed to an adult male, stands out. A partial skeleton (ER 1500), attributed to an adult female, has also been found at Koobi Fora. On the basis of these and other remains, it has been established that the height, weight, and body proportions of *Paranthropus boisei* were basically similar to those of the australopithecines (Figure 3.4).

In addition to those at Olduvai, the Omo River valley, and the shores of Lake Turkana, other East African deposits between 1.3 and 1.4 million years old have yielded fossils of *Paranthropus boisei*. In 1964 one of the most complete mandibles known of this species was found in the Peninj deposits in Tanzania (Peninj 1); in 1970 part of a skull (known as CH 1) was found at the Kenyan site of Chesowanja. In October 1997 a team led by Gen Suwa published the find, in the Ethiopian Konso deposit, of the only skull complete with mandible known for this species (KGA 10-525); it was dated at 1.4 million years old.

While the discovery of a skull together with its jawbone is very important, there is another aspect of the Konso find which is especially significant. The bovid fauna in this deposit consists of 80 percent alcelaphins (animals of the wildebeest type, such as damaliscus and topi), while fossils of equines are also frequent. Alcelaphins are a group (technically a tribe) of bovids which eat C4 plants (as do equines); as we have already seen, these form the predominant vegetation in grasslands. In other words, the fossil fauna indicates an open, dry environment of the savanna type.

As we have already noted, *Paranthropus* fossils have also been found in the far south of the continent; these are assigned to the species *Paranthropus robustus*. Broom's first finds in Kromdraai were supplemented by others made by Broom together with John Robinson ten years later, in 1948, in the neighboring Swartkrans cave.

On the basis of paleontological analysis, the stratigraphic levels containing fossil *Paranthropus* in Kromdraai and Swartkrans are dated between 1.8 (the oldest) and 1 million years (the most recent) old.

91

Figure 6.3   Male *Paranthropus boisei*

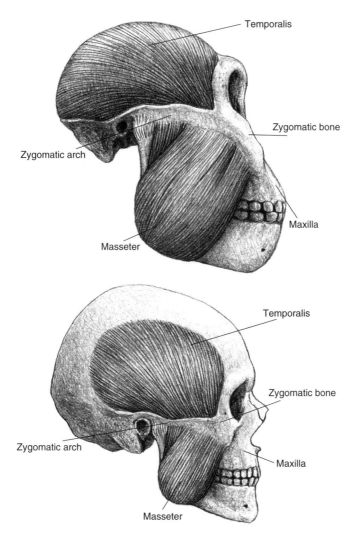

Figure 6.4   Bones and muscles involved in mastication. The temporalis muscle is fan-shaped. Its wider end originates on the external surface of the side walls of the cranium, while the narrower part passes through the temporal fossa to insert in the mandible. Contraction of these muscles can easily be felt by palpating the temples while clenching the teeth. The masseter muscles extend from the lower edge of each zygomatic arch to the outer faces of the mandibular ramus and the posterior parts of the mandibular body (see Figure 6.1). It is also easy to locate these muscles by touching the cheekbones while clenching the molars

Most of the fossils found in Kromdraai are fragments of skulls, jawbones, teeth, and a few fragments of arm, pelvis, and foot bones. The collection from Swartkrans is larger, and includes a fairly complete, though distorted skull (SK 48) as well as other skull and jawbone fragments and teeth. There are also numerous postcranial fragments, including arm, leg, and pelvis bones, and vertebrae. Nevertheless, the attribution of these postcranial remains to *Paranthropus* is problematic, given that fossils of a primitive representative of the genus *Homo* are found at the same levels.

To complete this account of the geographic distribution of *Paranthropus*, we should note that the great fossil "gap" between Tanzania and South Africa might begin to be filled with the finds now being made by a team led by Friedemann Schrenk and Tim Bromage, in deposits on the eastern shore of Lake Malawi (in Malawi). In addition to a jawbone attributed to the genus *Homo*, which we shall describe later, it is known that this team has found a fragmented upper jawbone of *Paranthropus*, details of which have yet to be published.

The Specialist

To the biologist, a *specialist* organism is one that has adapted to a very specific lifestyle (or ecological niche). This specialization will usually affect its diet, which in turn will be reflected in marked adaptations of its chewing and digestive apparatus. Conversely, organisms which maintain a more varied lifestyle and diet, and are thus less specialized, are known as *generalists*.

Specimens of *Paranthropus* show a series of unique traits which allow us to distinguish them easily from other hominids. These characteristics relate to their impressive masticatory (chewing) apparatus, which suggests marked specialization in diet among *Paranthropus*.

Mastication, or chewing, is the process whereby food is crushed by the pressure of teeth set in a mobile bone (the mandible or lower jawbone) against teeth set in a fixed bone (the maxilla or upper jawbone). As in a mortar, the food can be simply crushed, if the mandible moves only up and down, or ground, if this vertical movement is accompanied by other, horizontal movements. Both the pressure of the mandible on the maxilla (i.e., the strength of the bite) and the direction of the movements made depend on the muscles attached to the mandible.

The movements of the mandible during chewing are very complex, and involve many muscles. However, we can simplify the effects which interest us, noting that vertical movements of the mandible are generated basically by two groups of muscles located symmetrically at either side of the head: the temporalis muscles and the masseter muscles (Figure 6.4). The horizontal movements of the mandible, on the other hand, depend principally on the action of the lateral pterygoid muscles.

The three species of *Paranthropus* show different degrees of specialization of the masticatory apparatus, which is very marked in *Paranthropus boisei*, less pronounced in *Paranthropus robustus*, and only sketchily apparent in *Paranthropus aethiopicus*.

Beginning with dentition: the molars of *Paranthropus* have a fairly thick layer of enamel, thicker than that of any other hominid or ape. This characteristic is clearly related to a diet which causes heavy wear on the molars, either because the food contains hard particles or because it requires prolonged chewing, or both. Moreover, the surface of the teeth used to crush food is very large in relation to body weight. This is due to the increased size of the molars and particularly to the fact that the premolars are "molariform" – they are shaped like molars, substantially increasing the total area of teeth dedicated to crushing food. Parallel with the increase in the size of the molars and premolars, the anterior teeth of *Paranthropus*, the canines and incisors, are much reduced in size.

In apes the premolars are not "molariform," so the area of the teeth dedicated to crushing food is restricted to the molars (and the reader will recall that the incisors and canines are large). In *Paranthropus*, in contrast, the mastication zone has extended to the premolars, which in these species are "molariform."

The maxilla (upper jawbone) of *Paranthropus* is further back under the cranium than is usually seen in apes and australopithecines, and the premolar area is thus also further back. The zygomatic bones, on the other hand, are further forward, bringing the zone in which the masseter muscles originate further forward. These modifications increase the strength of the bite in the premolar area (Figure 6.4).

Furthermore, the force exerted by a muscle is directly related to the number of muscle fibers it contains – in other words, to its thickness. There is a physical limit on the size of the temporalis muscle: it has to pass through the temporal fossa, the diameter of which determines the maximum thickness of the muscle. In *Paranthropus* the zygomatic arch (which forms the lateral end of each of the temporal fossae) is strongly curved outward, markedly increasing the diameter of each fossa and indicating extremely thick and powerful temporalis muscles. These muscles were so big in *Paranthropus* that in males they result in the appearance of a sagittal ridge along the mid-line of the upper surface of the cranium.

In addition to the increase in size of the temporalis muscle, the mandibular ramus on each side is very long, increasing the power of the action of the temporalis and masseter muscles. This results in a corresponding increase in the height of the zygomatic bones and the maxilla.

The mandible (lower jawbone) in *Paranthropus* is very robust (that is, it has a very thick mandibular body), responding to the need for the bone to resist and dissipate the powerful stresses caused by the great force of chewing. In addition, a *Paranthropus* mandible is easily distinguished from that of any other hominid or ape because it is very wide in relation to its length. This form is ideal for circular movements of the mandible, used to grind hard food.

The effect of these skeletal modifications on the physiognomy (facial structure) of *Paranthropus* was considerable (Figure 6.3). As a result of the backward shift of the maxilla and the forward shift of the cheekbones, the face of *Paranthropus* was not prognathic, with the face projected forward, but flat or even concave. Moreover, the region of the cheekbones was much wider because of the lateral widening of the temporal fossae. Finally, the face was also very long, because of the increased length of the mandibular ramus and the zygomatic bones and maxilla. It is the unmistakable face of very specialized hominids.

7

A New Kind of Hominid

Whatever a man comes to be worth, he can never hope to exceed his innate worth as a man.

Antonio Machado, *Juan de Mairena*

The First Humans

The term *homo* ("man," in the generic sense of "human being") was used by Linnaeus in 1758 to denote the genus to which our own species (*Homo sapiens*) belongs. The terms "humanity" and "human" are usually restricted to representatives of our genus, so when we explore the origin and evolution of the genus *Homo* we are referring to the origin and evolution of humans.

Until the early 1990s the oldest fossils assigned to our genus consisted of an assemblage of isolated teeth (around 2.1 million years old) and jawbone fragments (around 2.5 million years old) found in the deposits of the Omo River valley. However, the allocation of these fossils to the genus *Homo* was disputed by many authors, who called for more solid evidence that *Homo* did in fact exist more than 2 million years ago. Those who believed that the genus was in existence at this time based their argument on the identification of stone tools in sediments about 2.3 million years old, also in the Omo deposits. But this was a debatable argument based on a premise which had yet to be proved: that the only hominids capable of manufacturing stone tools are members of the genus *Homo*.

Since the early 1990s, a series of fossils over 2 million years old, and attributed to our genus, has been presented to the scientific community. In 1992 it was reported that a fragment of temporal bone (known as BCI, and found in 1965 in a deposit close to Lake Baringo in Kenya) belonged to our own genus; this was dated at around 2.5 million years old. However, as we do not know exactly where it was found, its position in the stratigraphy of the deposit cannot be determined, and its age cannot be reliably established. Moreover, not all researchers accept that this fossil belongs to the genus *Homo*. In our view, the Lake Baringo fossil does belong to our genus, although it cannot be assigned to a particular species. A very complete mandible (UR 501), found by Schrenk and Bromage's team on the western shore of Lake Malawi, is also attributed to *Homo* (specifically, to *Homo rudolfensis*). This fossil is dated at between 2.3 and 2.5 million years old, on the basis of paleontological data; this age is considered approximate and could be reappraised. Again, the attribution of this mandible to *Homo* cannot be considered definitive, given that, as we understand it, it could be that of a *Paranthropus*.

In 1994 Johanson's team found a very complete maxilla (A.L. 666-1) in the Hadar region; the allocation of this find to the genus *Homo* seems clear. The fossil was found immediately below a volcanic tuff dated at around 2.3 million years old on the basis of radiometric techniques. Around twenty stone tools were found together with the human fossil; this represents the oldest known case of an association of stone tools with hominid remains. This find has reinforced the hypothesis that it was representatives of *Homo* who made the first tools. The oldest known worked tools were found in another deposit in the Hadar region, Gona, which is dated at around 2.5 million years old.

Systematic dressing of stone to obtain tools appears to have been one of the keys to our genus's ability to inhabit very varied environments and to gain access to new resources. It is therefore worth leaving bones aside for a moment to consider stones.

The Stone Cutters

Humans are not the only animals to use tools. Chimpanzees also do so, but their tools are natural objects suited to their function, such as the sticks they prepare by stripping away the side shoots and then insert into the openings of termites' nests to "fish" for the insects. In these cases the

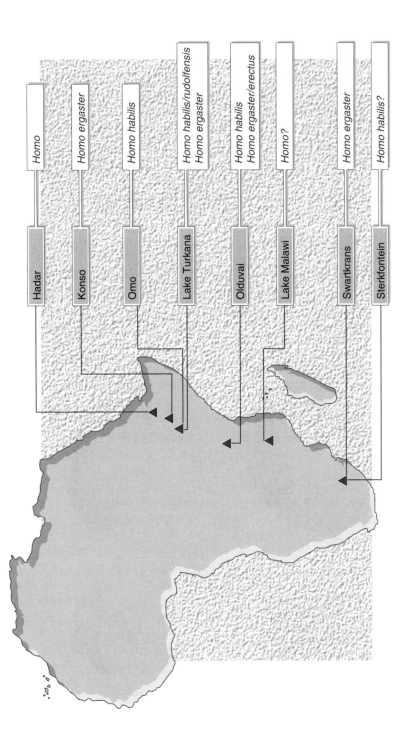

Figure 7.1 Location of deposits containing fossils of the first representatives of the genus *Homo*

Hadar — *Homo*

Konso — *Homo ergaster*

Omo — *Homo habilis*

Lake Turkana — *Homo habilis/rudolfensis*
Homo ergaster

Olduvai — *Homo habilis*
Homo ergaster/erectus

Lake Malawi — *Homo?*

Swartkrans — *Homo ergaster*

Sterkfontein — *Homo habilis?*

tool is already formed in nature and no great effort is required to visualize the final object that needs to be prepared.

Humans are the only primates who really produce tools on the basis of a shape that exists only in their minds, which they "impose" on the stone. It is said that the great Renaissance sculptor Michelangelo said of his sculptures, such as *David* and the *Pietà*, that they were already enclosed in the block of marble and all he did was to take away what was not needed. It was something along these lines, although on a more modest scale, that humans were doing when they began carving stone.

The first stone tools are known as Oldowan (a name derived from Olduvai), and they are very simple. They are classified as Mode 1 tools, and they consist of pebbles and rocks carved without any standardized form; they include what are known as *choppers* (worked on only one side), *chopping tools* (worked on two sides), and unworked flakes (Figure 7.2). The process of manufacturing these tools involves a short sequence of strokes. Unmodified pebbles and rocks were also used as hammers and anvils.

These simple stone tools would have given their makers something they lacked, owing to the reduction of their canine teeth: a cutting edge. But

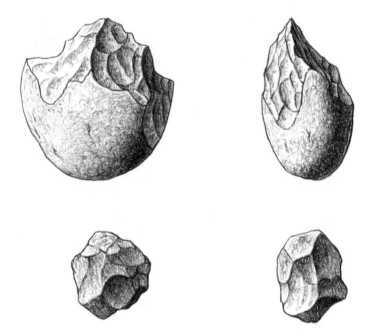

Figure 7.2 Some typical Mode 1 stone tools. *Top*: two views of a chopping tool; *below*: two views of a polyhedron

even making such simple tools is more difficult than it seems. Archeologist Nick Toth and psychologist Sue Savage-Rumbaugh conducted an experiment to teach a male bonobo (pygmy chimpanzee) named Kanzi, who has shown remarkable capacity for manipulating symbols, to carve stone. (We shall return to Kanzi later, in the chapter on language.)

In front of Kanzi, Nick Toth struck a pebble with another to obtain flakes; with the edge of the flakes he then cut a string and opened a box containing food. Kanzi immediately understood the usefulness of the flakes and learned to select the best ones and to use them, but when it came to producing them all his efforts were in vain. There is no doubt that Kanzi has in his mind the idea of the flake he wants to obtain, but he is unable to judge how to strike one stone against the other. The reader may also try to do it: you will see that it is not so easy, particularly if you value your fingers! In order to carve it, the stone needs to be held steady with one hand and struck on its edge, slightly at a tangent, with another stone which acts as a hammer or mallet. Flakes will come away from the stone which is struck, leaving a carved core. Both the flakes and the core can be used as tools, because they have edges. Nevertheless, Kanzi discovered that by throwing a stone against the ground or against a rock the stone would break apart and produce usable edges, although the result did not resemble prehistoric tools.

As we have already noted, chimpanzees' hands are not very well adapted to manipulating small objects with precision, because of the length of the palm and all the fingers except for the thumb, the tip of which is some distance away from the tips of the other fingers. It is a hand made for hanging from branches, not for carving. However, it is likely that the main problem lies in the inability to coordinate the necessary movements of the arms, wrists, and hands.

The australopithecines, at least since *Australopithecus afarensis*, have hands which are already very similar to our own, and we may therefore suppose that they had the mental capacity (just as chimpanzees do) and the necessary coordination to produce flakes. If they did not do so, it is perhaps because they did not need sharpened edges to cut with. It may be that this need arose when the first humans began to eat meat, and required edges both to cut the thick skin of large animals and to cut tendons and slice up muscle. They also used stones to break open bones and extract the marrow, although this activity is similar to that practiced by chimpanzees when they open nuts using a stone as a hammer and another, fixed on the ground, as an anvil. The animals that humans ate would not necessarily have been hunted by them: often they would have been carrion. They probably also used the sharp edges of stone tools to cut plants, and some believe that what *Paranthropus* did to prepare tough plant material using their hyperdeveloped chewing apparatus, the earliest *Homo* did with their stone tools.

The Diversification of *Homo*

The fossil record of *Homo* from deposits less than 2 million years old is much richer and more varied. Until a few years ago, all the African fossils dated between 2 and 1.4 million years old were attributed to two species. The oldest and most primitive were assigned to the species *Homo habilis*, while the more recent and more evolved were seen as African representatives of *Homo erectus*, a species defined on the basis of fossils found in Java and China (which we shall consider later). However, since the 1980s the discovery of fossils with different morphology has led scientists to rethink the number of species represented by African fossils over one million years old.

Bernard Wood, who studied the cranial remains and fossil teeth from Koobi Fora, has been the main agent of this revision. In his opinion, three species can be distinguished among the oldest human fossils of Africa: *Homo habilis*, *Homo rudolfensis* (the name *rudolfensis* refers to the old name for Lake Turkana, Lake Rudolf), and *Homo ergaster* (*ergaster* meaning "worker" in Greek). The fossils previously attributed to *Homo habilis* are divided between the first two species, while the latter includes virtually all the remains initially assigned to *Homo erectus* (with the exception of skull OH 9, of which more later).

According to Wood, *Homo rudolfensis* is distinguished by the combination of a large brain with highly developed masticatory apparatus, similar in some respects to that of *Paranthropus*. *Homo habilis*, on the other hand, had a brain somewhat larger than that of australopithecines and *Paranthropus*, but smaller than that of *Homo rudolfensis*, combined with a primitive overall architecture of the skull very similar to that of *Australopithecus africanus*. It was distinguished from the latter by a less developed and shorter masticatory apparatus, the smaller size of the molars and the presence of a brow ridge separated from the rest of the frontal bone by a shallow furrow (Figure 7.3).

Nevertheless, in our opinion the main fossils used to justify the existence of *Homo rudolfensis* could easily be included in *Homo habilis* if we accept that the species showed a marked sexual dimorphism, like the australopithecines and *Paranthropus*.

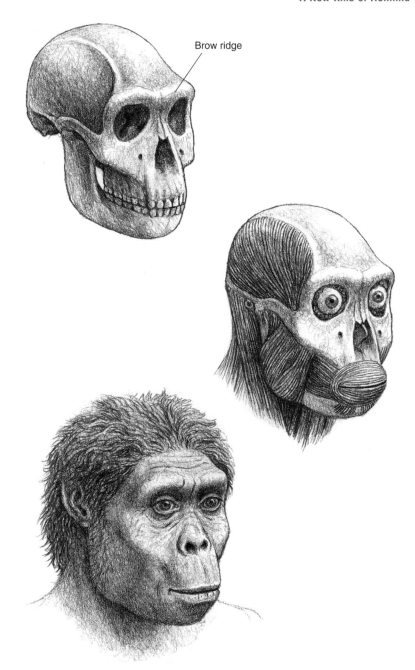

Brow ridge

Figure 7.3   Female *Homo habilis*

Let us move on, continuing the assignation of fossils to the species proposed by Wood in order to complete our account of the information available.

Homo rudolfensis is a species with a chronological range extending from 1.9 to 1.6 million years ago, and of which the only fossils known come from deposits on the shores of Lake Turkana, particularly those of Koobi Fora. The most significant fossils for this species are skulls ER 1590, ER 3732, and ER 1470. The most complete of these is ER 1470, which was found by Richard Leakey's team in 1972. This fossil has been dated at 1.9 million years old, and it is the holotype of the species.

The first fossils attributed to the species *Homo habilis* were found on November 2, 1960, during the course of Louis and Mary Leakey's excavations in Olduvai Gorge. They consisted of part of a mandible, cranial bones, and bones from the hand of a juvenile. These remains were numbered as OH 7 and were chosen as the holotype of the species when Louis Leakey, Phillip Tobias, and John Napier (1917–87) created it in 1964. Since the first finds, the collection of remains of *Homo habilis* from Olduvai has increased substantially, and it now covers the period between 1.8 and 1.6 million years ago. The skull remains include an almost complete but very flattened skull known as OH 24 (nicknamed Twiggy), and an assemblage of skull fragments with mandible numbered as OH 13 (known as Cinderella or Cindy).

The Olduvai collection also includes a good representation of the postcranial skeleton, including specimen OH 62, found by Johanson and White's team in 1986, which consists of part of the skeleton of an adult female. Study of the arm and leg bones of OH 62 produced surprising results. First, estimates of the individual's height gave a figure of barely one meter, making it the smallest of all the hominid fossils (including Lucy). Second, the ratio between the length of humerus and femur calculated for OH 62 is 95 percent, closer to that of chimpanzees than that estimated for Lucy herself (85 percent).

Richard Leakey's excavations at Koobi Fora have yielded, since the early 1970s, the largest and most complete collection of *Homo habilis* fossils currently available. The collection includes, in addition to many more fragmented skull remains, some very complete skulls (ER 1805 and ER 1813), and abundant representation of jawbone fragments and isolated teeth. Abundant postcranial skeleton fossils have been recovered, including an incomplete skeleton (ER 3735) which is similar to OH 62. The Koobi Fora fossils are concentrated within a short period from approximately 1.9 to around 1.8 million years ago.

In contrast to the rich collection of *Homo habilis* fossils from Koobi Fora, the remains of this species found in the Omo River deposits are limited to isolated teeth and very fragmentary remains which are difficult to assign to species. One exception is part of a skull (L 894-1) which is around 1.9 million years old.

To complete this overview, we have one problematic fossil from the South African Sterkfontein deposit, which is about 1.8 million years old (it comes from a stratigraphic level more recent than that containing fossils of *Australopithecus africanus*). This is a fragmentary skull (Stw 53) which some researchers assign to *Homo habilis*, while others see it as belonging to a species of *Homo* which has not yet been named.

Ready for the Great Leap

Homo ergaster is distinguished from the preceding *Homo* species by a marked increase in brain size, the presence of a brow ridge clearly separated from the rest of the frontal bone by a pronounced furrow, a positioning of the nasal bones such that the nose projects in profile, and a shortening of the facial skeleton and reduction in the relative size of the molars (Figure 7.4).

The fossils assigned to *Homo ergaster* cover the period between 1.8 and 1.4 million years ago. The oldest are from Koobi Fora, and the most recent could be a mandible from the Konso deposit, discovered in the same geological level as a set of stone artifacts and the remains of the *Paranthropus boisei* we discussed above.

The most notable fossil attributed to *Homo ergaster* is the skeleton found by Richard Leakey's team in 1984 on the bank of the Nariokotome River, a few kilometers west of Lake Turkana (this is officially numbered WT 15000, but it is familiarly known as Turkana Boy). This skeleton is around 1.5 million years old, and most of the bones of the body are preserved, including the cranium and the mandible. This discovery is one of the most important in the history of paleoanthropology, because it is the most complete skeleton of a hominid of this antiquity. All that is missing are the bones of the feet and almost all the bones of the hands. On the basis of the state of development of the bones and teeth, the skeleton has been identified as a child of between nine and ten years old; the morphology of the pelvic bones indicates a male. Because the skeleton is so complete, it has been possible to make a very reliable estimate of his height. The result is surprising, since it gives a figure of around 1.60 m, suggesting that he

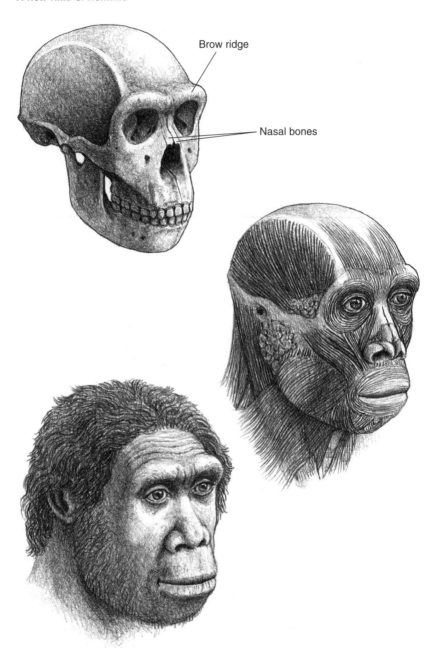

Brow ridge

Nasal bones

Figure 7.4   Male *Homo ergaster*

would have been more than 1.80 m tall when he finished growing. Another very interesting feature which can be reliably established for WT 15000 is the ratio between the lengths of the humerus and femur (Figure 7.5). This ratio is 74 percent, entirely human and contrasting with that estimated for the *Homo habilis* fossil OH 62 (95 percent).

Two very complete skulls known as ER 3733 and ER 3883 were found at Koobi Fora. The first of these was found in 1975, in sediments around 1.8 million years old; the second was found a year later in another, more recent stratum, around 1.6 million years old. Femurs ER 1472 and ER 1481, and coxal bone ER 3228, found at Koobi Fora, and coxal bone OH 28 from Olduvai, may also belong to *Homo ergaster*.

Outside of East Africa *Homo ergaster* is represented by an assemblage of fossils recovered in the South African deposit of Swartkrans (in the same sediments which contained fossils of *Paranthropus robustus*, about 1.8 million years old); these include remains of jawbones, teeth, and part of a skull (SK 847).

Homo ergaster appears as the most human of the early species of *Homo*. In addition to its large brain, this species exhibits a height and limb proportions similar to those of later humans. Moreover, shortly after *Homo ergaster* emerged, a new form of stone-dressing appeared, known as the Acheulian industry.

The Acheulian, or Mode 2 industry, much more elaborate than the Oldowan or Mode 1 culture, includes cores or large flakes worked on two sides, known as *bifaces* and including handaxes, cleavers, and picks. These tools show a marked degree of standardization in manufacture, and they require a long sequence of gestures to make them, including turning the core in the hand while continuing to strike it with the hammer to obtain flakes. The result is a tool in which all or almost all of its edge forms a blade. Handaxes are symmetrical bifaces with lateral cutting edges which converge to a pointed tip; cleavers are bifaces with a straight edge on one end (Figure 7.6). The handaxes were probably multipurpose tools, used to cut meat, work wood, and perhaps also to prepare skins. The oldest Acheulian industry known is around 1.6 million years old, and was found in the Olduvai deposit in Tanzania.

Family Relationships

One of the most important aspects of a paleontologist's work is the attempt to reconstruct the chain of ancestors and descendants who

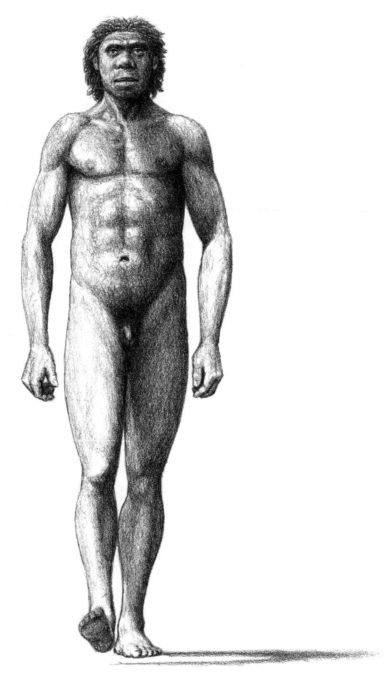

Figure 7.5   *Homo ergaster*

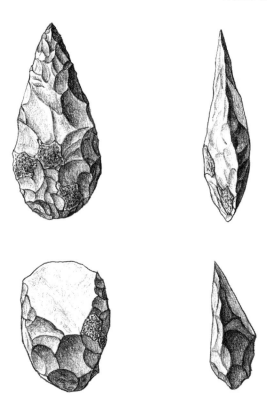

Figure 7.6 Typical Mode 2 tools. *Top*: two views of a handax; *below*: two views of a cleaver

make up the Story of Life. Until we have a time machine to take us back into the past, no hypothesis of ancestor–descendant relationship can be confirmed. Nevertheless, it is possible to continue reducing the number of likely hypotheses, discarding others which are incompatible with three types of data on the species represented by fossils: their age, their geographical distribution, and their degree of evolutionary relationship.

It is clear that more recent species cannot be proposed as ancestors of older species, nor the latter as descendants of the former. Nor can we assume an ancestor–descendant relationship between species which are strictly contemporary with one another.

Moreover, in order for a species to be accepted as the ancestor of another, it is not enough that it be older: it is also essential that it lived in the same geographical region in which the assumed descendant species originated.

Another criterion used by paleontologists when deciding whether to retain or discard their hypotheses is the degree of evolutionary relationship between the species. The closer the evolutionary relationship, the more likely it is that there is an ancestor–descendant relationship between them: the parents of a person can only be found among that individual's first-degree relations, and all we need to do is set aside any siblings. If we knew the degree of evolutionary relationship between all the species of the past, the problem would be reduced to distinguishing between "parent" and "sibling" species, and this could be done very easily if we also knew the age and geographical distribution of the species in question.

However, the reality is more complicated, and we have already seen in previous chapters that it is not always possible to date fossils precisely. Similarly, we cannot be sure that new fossils will not appear, broadening the chronological range or area of distribution of a species. Even so, the thorniest problem for the paleontologist is establishing degrees of evolutionary relationship. At the end of the day, the temporal and geographical distribution of the species of the past can be established directly on the basis of fossils; evolutionary relationships, however, are not obvious, but have to be deduced by the scientists.

The Science of Relationships

Evolutionary relationships are technically known as *phylogenetic relationships*. In order to deduce these relationships from the fossil record, paleontologists use a method based on very simple principles, implicit in the Darwinist theory of lines of descent with modification, but much more complex in application. In essence, the process involves tracking relationship on the basis of similarity between the different species: the greater the similarity, the closer the evolutionary relationship. But not all resemblances are the same. Apart from pure coincidence (such as the fact that spiders and octopuses have the same number of extremities), similarity may be due to two main reasons: adaptive convergence (analogy), or inheritance from a common ancestor (homology).

Adaptive convergence occurs when two anatomical structures resemble one another not because they were inherited from a common ancestor, but because they perform similar functions and natural selection has shaped them in similar ways. One example of this is the fins of sharks and the flippers of dolphins: they resemble each other because they have a similar function, but their anatomical structure is very different.

Similarities due to *inheritance from a common ancestor*, on the other hand, maintain their structural resemblance even though they may show superficial differences if the organs perform different functions. This may be seen if we compare the wing of a bat with a dolphin's flipper, where the differences are due to the difference in function, but the anatomical organization is very similar.

In the examples given above, the scientist would reject the superficial resemblance between the shark's fin and the dolphin's flipper, and on the basis of the structural similarity between the wing of the bat and the dolphin's flipper would suggest that the two share a common ancestor (from whom they inherited the structure of their extremities), and that this was not the ancestor of the shark; thus, the bat and the dolphin are more closely related.

This is the first step a paleontologist must take when establishing relationships between species: distinguishing between common inheritance characters and adaptive convergences. However, common inheritance characters are themselves not all the same, nor do they all hold equal value in reconstructing phylogenetic relationships. We need to distinguish between *primitive characters* and *derived* or *evolved characters*. The former do not give us any information on the degree of relationship; only the latter are useful in establishing whether the relationship between species is close or more distant. The more derived traits two species share, the closer their evolutionary relationship.

A good example of this way of working will be found in the chapter looking at the characteristics of different groups of primates. In that chapter, we defined the primates with derived traits such as opposable big toes or the presence of flat nails, both inherited from the exclusive common ancestor of the primates. At the same time, we eliminate other characteristics such as the possession of clavicles or five fingers and five toes, since these are primitive traits of all mammals (and do not indicate the existence of an exclusive common ancestor for primates). Other derived traits, such as the absence of a hairless snout, the loss of a premolar, and adaptations to brachiation, are used to distinguish haplorrhines, catarrhines, and hominoids respectively.

Thus the task of the researcher into phylogenetic relationships can be summed up in two stages. First, distinguishing the convergent characters from common inheritance characters, and separating primitive from derived traits. Then, the degree of evolutionary relationship between the species in question can be established on the basis of the presence of derived traits.

This is not always an easy task, given the complexity of the evolutionary process and above all the scarcity and state of preservation of the fossils themselves. These difficulties mean that different researchers arrive at different conclusions on the nature of the characters used in the analysis. As a result, there is no consensus on the hominid evolutionary tree, and various hypotheses have been put forward by different authors.

The Hominid Tree

If we were to describe all the evolutionary trees put forward for hominids, explaining how the appearance of new fossils has progressively eliminated some and resulted in the presentation of other new theories, we would no doubt try the patience of our readers. We therefore prefer to discuss the evolutionary tree which in our view is the most compatible with the evidence currently available (Figure 7.7).

In view of its antiquity and the primitive nature of the majority of its traits, the species *Ardipithecus ramidus* appears the ideal candidate for the

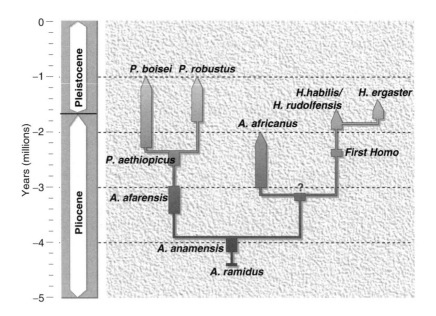

Figure 7.7   The authors' evolutionary diagram for the first hominids

role of ancestor of the other hominids. Until we have information on this species' mode of locomotion, the only derived hominid characteristic found in *Ardipithecus ramidus* appears to be the morphology of the upper canine and the third lower premolar.

More recent than *Ardipithecus ramidus*, *Australopithecus anamensis* shows at least one other derived trait: the thick layer of enamel on the molars. (If *Ardipithecus ramidus* is eventually shown not to have been bipedal, bipedalism would be another new feature in *Australopithecus anamensis*.) We do not yet know enough about the anatomy of *Australopithecus bahrelghazali* to be able to include it with confidence in the hominid evolutionary tree.

The majority of authors consider *Australopithecus afarensis* to be the last common ancestor of later hominids (Figure 7.7). However, we take a different view. Our studies on the region at the base of the skull related to mastication have led us to conclude that the chewing apparatus of *Australopithecus afarensis* was proportionately wider than that of apes, australopithecines, and humans, but equivalent to that of *Paranthropus*. The widening of the masticatory apparatus, which is reflected in the presence of relatively wide jawbones, is characteristic of *Paranthropus* and tends to favor circular movements of the mandible, necessary for processing tough foodstuffs. In our opinion, this resemblance between *Australopithecus afarensis* and *Paranthropus* is not a phenomenon of adaptive convergence but a shared derived character. Consequently, we believe that *Australopithecus afarensis* already formed part of the lineage of *Paranthropus* and could not have been a direct ancestor of the *Homo* line (Figure 7.7).

The *Paranthropus* branch is characterized by a number of derived traits, mainly related to the particular specialization of its masticatory apparatus. Nevertheless, there is a gradation within this specialization which helps us to establish the relationship between the three known species of *Paranthropus*.

Thus, in aspects such as its high degree of prognathism (compare Figure 6.3 and Figure 6.4), and a number of details of the morphology of the base of the skull, *Paranthropus aethiopicus* appears to be the least specialized of the species. This more primitive cranial anatomy, combined with its greater age, makes this species the ideal candidate to be considered the ancestor of the other *Paranthropus* species.

The first representatives of the genus *Homo* are very similar to the specimens of *Australopithecus africanus*, from which they are distinguished by their larger brain, somewhat less developed masticatory apparatus and the presence of a slight brow ridge. Apart from these differences, the architecture and morphology of the rest of the cranium are very

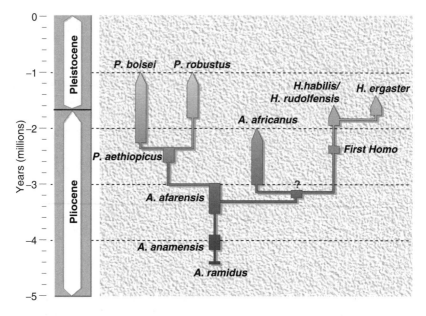

Figure 7.8 Alternative evolutionary diagram for the first hominids

similar in the two species (compare Figure 4.4 and Figure 7.4). This seems to us to indicate a very close relationship between *Australopithecus africanus* and *Homo*. In our view, the common ancestor of the two lived between 4 and 3 million years ago.

As far as evolutionary relationships between species in our own genus are concerned, the question of whether *Homo rudolfensis* existed or not makes very little difference. If we recognize only the existence of *Homo habilis* the problem is somewhat simpler, since only two species remain to be related, and on the basis of both chronology and cranial anatomy *Homo habilis* could be the ancestor of *Homo ergaster*.

In any case, if we agree the existence of *Homo rudolfensis*, the species has to be considered an evolutionary offshoot within the *Homo* genus, since the anatomy of its masticatory apparatus is too specialized to propose it as the ancestor of the other species of the genus. The position of *Homo habilis* as the original species of *Homo* would thus remain unchallenged.

Our evolutionary tree for hominids implies that *Paranthropus* originated in East Africa. From there, representatives of this lineage reached the ecosystems of the south of the continent, where they evolved separately from their East African fellows and gave rise to a different species,

114

Paranthropus robustus. This phenomenon of local evolution of a peripheral population has been common in the history of hominids, and we will encounter it again later on in the evolution of *Homo.*

Since the oldest fossils of our genus come from the east of the continent, the simplest hypothesis is that it was there too that our genus originated, and from there spread to the more southerly regions of Africa.

Now that we have described the fossil record of the oldest hominids, examined their basic adaptations and discussed their family relationships, it is time to look at other questions of their life, such as the size of their brains, their diet, their development, sexuality, and sociability.

8

The Evolution of the Brain

I thence concluded that I was a substance whose whole essence or nature consists only in thinking.

René Descartes, *Discourse on Method*

The Organ of Intelligence

Human beings are characterized by a much more developed intelligence than other animals. Although what we call "intelligence" is a concept that is difficult to define and very problematic in terms of measurement, it is clear that it is related to certain skills which are unique to us. No other animal has yet proved capable of writing a book, composing a symphony, traveling to the moon, designing and constructing a computer, or simply wondering about its origin and its destiny.

The region of our organism devoted to performing these functions is the brain, which is housed in the interior of the cranial cavity and is composed of three organs, the cerebrum, the cerebellum, and the brainstem (Figure 8.1). In very general terms, we can say that the superior functions related to intelligence (for example the capacity for abstraction, the association of information, or the ability to encode and decode information via an articulated language) are performed in the cerebrum, which is divided into two hemispheres, left and right. The cerebellum is responsible for general motor coordination and balance, while the brainstem acts as our "automatic pilot," responsible for maintaining regular functions such as breathing and heartbeat.

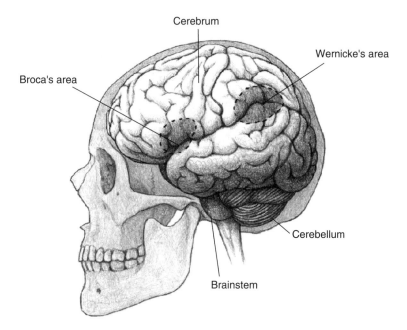

Figure 8.1   The human brain The areas of the cerebral cortex directly related to language are shaded

Research into the evolution of the brain in hominids has focused on two main areas: evaluating the increase in the size of the brain itself over the course of hominid evolution (the process of encephalization), and analyzing the morphology of the organs of the brain, especially the cerebrum.

Large and Small Brains

When, in our lectures, we ask which is the animal species with the largest brain, almost all students immediately reply: humans. The conviction that, in line with our superior intelligence, our species is the most encephalized in the animal kingdom, and that this is the basic characteristic that distinguishes us from other animals, is embedded in our culture. However, this is not the case, or at least not in absolute terms.

The average size (mass) of the human brain is around 1,250 g, and although this is greater than that of any other primate species (chimpanzees and gorillas have average brain mass of around 400 and 500 g, respectively), and of the majority of animals, it is markedly less than that of the large mammals, whose brains are much larger than ours: the blue whale (the largest animal that has ever existed) has a brain mass of around 6,800 g, while the African elephant (the largest living land mammal) has a brain mass of about 5,700 g.

Does this mean that blue whales and elephants are more encephalized, and therefore more intelligent than humans? Of course not. What happens is that since the work of the brain is to coordinate the functions of the rest of the body, larger bodies predictably require larger brains to ensure sufficient coordination. Thus when we compare the brain size of two species we have to consider the influence of body size on the size of the brain.

A simple way of relating body size to brain size is to divide brain mass by body mass; in this way we can find the ratio between the two. The greater the ratio, the more brain an animal would have per unit of body mass. On first view, this might seem the most appropriate way of comparing the brain size of animals of different sizes. However, this method also fails to reveal humans as the most encephalized species in the animal kingdom. Paradoxically, we are overtaken by the smallest mammals, which have proportionately larger brains.

Should we then resign ourselves to accepting that the smallest mammals are more encephalized than we are? We can rest easy: shrews do not have a more developed brain than we do. The truth is that brain size increases more slowly than body size: in other words, the size of the brain becomes proportionately smaller in larger mammals. This phenomenon, the change in the proportions of the organs as body size increases, is very common in living beings, and was discovered in the 1920s; it was named *allometry*, or the Law of Disharmony.

Allometry may occur during the growth of an organism: thus, for example, a newborn has a proportionately bigger head than an adult, and its limbs are relatively shorter. If a baby maintained these proportions as it grew, the result would be an adult of very alarming appearance (Figure 8.2).

Allometry may also occur between adults of the same species. A good example of this is the variation in "ideal weight" in relation to height. As height increases, the ideal weight becomes relatively greater: for a woman 150 cm tall the "ideal weight" is 47 kg (315 g per centimeter), while another woman 180 cm tall has an "ideal weight" of 67 kg (372 g per centimeter). But the kind of allometry which most interests us is the variation between species in the same group. Thus in mammals, brain size varies in

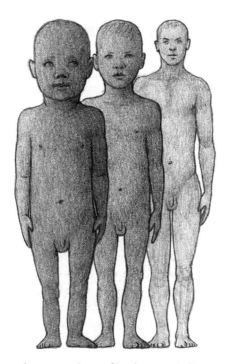

Figure 8.2   Change in proportions of body parts (*allometry*) during growth.
Left: a newborn; *center*: a three-year-old child; *right*: an adult – all drawn the same height

relation to body size, and as we have already noted, brain size becomes proportionately smaller as body weight increases. While in chimpanzees the ratio between brain mass and body mass averages 0.01 (one gram of brain per 100 g of body weight), in the mouse lemur (a small primate from Madagascar with a body mass of only 60 g and brain mass of 1.8 g) the ratio is around 0.03 (i.e. 3 g of brain for each 100 g of body weight).

Given that mammals' brains grow allometrically in relation to body size, the only way of comparing the brains of different species is to calculate the brain mass that each should have in relation to its body mass, and to compare these "ideal brain masses" with their actual brain mass. The species with the greatest surplus brain mass would be the most encephalized.

Clearly, the key to comparing species of different size lies in calculating the expected values of the brain mass of each. To this end the mathematical law which relates the body mass of an organism to its "ideal brain mass" has been calculated, on the basis of analysis of extensive data on brain mass and body mass in a large number of mammal species.

World Champions of Encephalization

The brain mass which would correspond to an organism's body mass is the *expected value* (its "ideal brain mass"). We can call the real size of its brain the *actual value*. The ratio of the two values (actual/expected) is known as the *index of encephalization*, and it measures the disparity between the size of brain an animal should have and the actual size of its brain.

When the index of encephalization for a species is equal to 1, we can say that its expected and actual values are equal and it therefore has the brain which corresponds to its body size. If the index is less than 1, the actual value is less than that expected and the species has a smaller brain than it should have according to its body mass, while an index higher than 1 indicates a brain larger than expected.

Various authors have calculated indices of encephalization for a large number of mammal species, and their results agree on a number of points.

First, primates appear as a highly encephalized group of mammals (but not the most encephalized, contrary to popular belief). In virtually all the primates which have been studied, the value calculated for the index of encephalization is greater than 1.

The second striking result is the fact that *Homo sapiens* is the most encephalized of all mammals, with an index of more than 7. In other words, our brain is more than seven times the size that could be expected in a mammal of our body mass.

Finally, one further singular fact: the species shown to be closest to humans in terms of encephalization are not primates but cetaceans (whales and dolphins), in particular dolphins, which have indices of encephalization above 4.

These results bear out the generalized belief that humans are the most encephalized species of the animal kingdom. But is this a characteristic common to all hominids?

Weighing Ghosts

Although the brain does not become fossilized, the skull, which contains the cavity which houses it, does. In skulls where the whole or a good part of the neurocranium is preserved, it is possible to measure the volume of the cavity which houses the brain. Since the brain occupies virtually the whole of this space, the volume of the cranial cavity (cranial capacity) can be used as a measure of the volume of the brain itself. Experimental studies on primates and humans have shown that cranial capacity is almost directly equivalent to brain mass: a capacity of 1,000 cc corresponds to a mass of 971 g.

Estimating body mass from fossils is a different matter. Here the problem is more complicated because there is no measurement we can take from the skeleton which would directly correspond to the individual's body mass. In tackling this problem, paleontologists start from an obvious fact: the size of the bones is related to the size of the body itself, so large bones should correspond to similarly large individuals.

However, not all the bones of the body follow this rule. For example, we have seen that *Paranthropus* had a highly developed jawbone, owing to its particular adaptation to diet. So if we applied the rule that large bones (and teeth) reflect large body size to the jawbones of *Paranthropus*, we would have to conclude that these hominids had a body size close to that of a female gorilla. This mistake was in fact made in the past, and is the source of the adjective "robust" usually given to *Paranthropus*.

In order to arrive at reliable estimates of the body mass of fossil hominids, we need to use bones or parts of bones which are clearly involved in supporting body weight. Modern humans have several bones which satisfy this requirement. In our species the weight of the trunk, the upper limbs and the head is transmitted through the lumbar region of the spine to the sacrum, in the pelvis. From here it passes alternately at each stride to each of the femurs (which also have to support the weight of the leg which is in the air), and through them down to the bones of the feet (which support virtually the entire body weight at each step). The lumbar and sacral vertebrae, the coxal bones and the bones of the lower limbs are thus good candidates for estimating an individual's body mass.

Nor are all the dimensions of a bone equally suitable for estimating body mass. For example, the length of the femur has more to do with an individual's height than his body mass. Since the body weight is transmitted from one bone to another through the joint surfaces of the two bones, these are directly related to body mass. The pressure exerted on a joint is directly proportional to body mass and inversely proportional to the joint's surface area; therefore, in order to avoid the pressure on a joint increasing with an increase in weight (and consequent injury to cartilage and bone tissue), a correlative increase in surface area is required. Thus, large joint surfaces indicate high body mass.

Having identified the bones and the anatomical regions which are suitable for calculating body mass, there is still another problem to resolve. What is the mathematical relationship between the joint surfaces of each bone and the body mass?

This question can be approached by analyzing data on joint surfaces and body mass in living primates, in order to determine the mathematical function relating them. But we have already seen that not all primates walk in the same way, nor do they have the same body proportions, and these factors affect how the body weight is transmitted. For example, in apes, which support themselves on four limbs when they move over the ground, the humerus is more involved in the transmission of weight, while the femur supports less weight than in modern humans. Given that the first hominids were fully bipedal, comparison with our own species seems most appropriate. Nevertheless, as we have noted in previous chapters, the body proportions of australopithecines and *Paranthropus* appear to have been intermediate between those of apes and our own. Thus the data on these primates also needs to be taken into account when estimating the body mass of fossil hominids.

Paleontologists meet a further obstacle when estimating the weight of bygone hominids: the scarcity of the fossil record itself. Few even roughly complete skeletons of the first hominids are known, and only in one or two cases has it proved possible to deduce an individual's mass on the basis of a range of bones (allowing us to refine the calculation). We also have very few bones suitable for use for each species, and therefore a very limited number of estimates for any one species. One final added difficulty is the fact that it is not always easy to assign a bone of the postcranial skeleton, found in an isolated position, to a specific species of hominid.

Despite all these problems, researchers such as Henry McHenry and William Jungers have undertaken a series of investigations which, using the available evidence and a variety of statistical techniques (together with a good dose of common sense), have yielded comparable results.

Australopithecines and *Paranthropus* appear to have had very similar body mass. Let us recall that the average body mass of male and female *Australopithecus afarensis*, the species for which body mass has been most accurately estimated, was 45 kg and 30 kg, respectively. *Homo habilis* does not appear to have been any bigger than the australopithecines and *Paranthropus*, whereas the values estimated for *Homo ergaster* are over 55 kg, indicating a marked increase in body mass toward values similar to those of our species.

The Brain Size of Fossil Hominids

Having estimated body mass, all we need to know, in order to calculate the indices of encephalization of fossil hominids, is their brain mass. The average brain mass[1] in the various species of hominid is as follows: *Australopithecus afarensis*, 426 g; *Australopithecus africanus*, 436 g; *Paranthropus robustus*, 523 g; *Paranthropus boisei*, 508 g; *Homo habilis/Homo rudolfensis*, 619 g; *Homo ergaster*, 805 g.

Using all of this, we can calculate the indices of encephalization of the first hominids. These are as follows: *Australopithecus afarensis*, 1.3; *Australopithecus africanus*, 1.4; *Paranthropus boisei*, 1.5; *Homo habilis/rudolfensis*, 1.8; *Homo ergaster*, 1.9. The sample for *Paranthropus robustus* is too small to make a satisfactorily reliable calculation of the index of encephalization.

On the basis of these values, we can draw some very interesting conclusions. First, australopithecines and *Paranthropus* show an index of encephalization higher than that of chimpanzees (1.2), but markedly lower than that of the first representatives of *Homo*, which have indices of encephalization 50 percent higher than that of chimpanzees and almost two-thirds that of modern humans (2.9).[2] Second, the increase in brain size noted in *Homo ergaster* compared to *Homo habilis* was compensated by a proportionate increase in body size, resulting in very similar indices of encephalization for the two species.

---

[1] Although we can calculate an average brain mass for each of the different species, the ranges of variation may overlap, with the higher values in one species overlapping with the lower values for another. This is the case, for example, with *Paranthropus boisei* and *Homo habilis*: the former has an average brain mass of 508 g, but a range of variation from 470 g (fossil ER 13750) to 590 g (Konso specimen). *Homo habilis*, on the other hand, has an average of 619 g but includes specimens with a brain mass of only 503 g (such as ER 1813) and others up to 661 g (OH 7), or 736 g, if fossil ER 1470 is included in this species (as it is for the purpose of our calculations).

[2] Our calculation of indices of encephalization was based on a different formula from that used in the section where we compared the encephalization of all mammals, which gave a value of 7 for our species. For this calculation we used the formula which estimates ideal brain mass on the basis of data from haplorrhine primates. This new calculation gives a value of 2.9 for our species. This means that we have a brain mass seven times the expected value for a mammal of our body mass, but only 2.9 times greater than that which would correspond to a haplorrhine primate of our size. Remember that primates are highly encephalized mammals.

Thus a marked increase in encephalization occurs around the origin of the genus *Homo*. But can this quantitative leap alone account for the greater intellectual complexity of the first humans, especially from *Homo ergaster* onward? Were the brains of the first humans simply bigger than those of australopithecines and *Paranthropus*, or were they also different in their organization? In order to answer these questions, we need to take a look at the part of the brain responsible for what we call intelligence: the cerebrum.

Surface Area of the Brain

The brain is made up of two different substances: white matter, which forms the internal part of the brain and the majority of its volume, and gray matter, which is almost entirely confined to a thin superficial layer, or cortex. Gray matter is made up of the bodies of the neurons (the highly specialized cells which form the nerve tissue), while white matter consists of the extension of these neurons (axons) which serve to connect and relate the neurons.

The cerebral cortex of non-mammalian vertebrates is very limited in area, and deals only with the analysis of olfactory stimulation; it is known as the *paleocortex*. In mammals the cerebrum is more highly developed and a new zone, the *neocortex*, appears on top of the paleocortex. In apes and humans the neocortex makes up virtually all of the cerebral cortex.

The surface of our brain is not smooth; it has a very complex topography in which a series of furrows (in fact folds in the cerebral cortex), known as *sulci*, can be seen; these run around mounds known as *gyri*. When the sulci are very marked they are known as fissures; these separate broad zones of the brain, or lobes. There are four lobes in each cerebral hemisphere: frontal, temporal, parietal, and occipital. The lobes are located in roughly the same region as the bones which bear the same names (Figure 8.3).

The brains of humans and apes are specialized in function, and we can locate regions associated with specific tasks in the surface of the brain. These functional maps can be related to the surface topography of sulci, gyri, fissures, and lobes, so that by examining the brain surface we can analyze the relative degree of development of each specific functional area and attempt to relate this to the skills and behavior of the animal.

Although there are numerous morphological differences between the human brain and that of apes, owing to their different kinds of mental

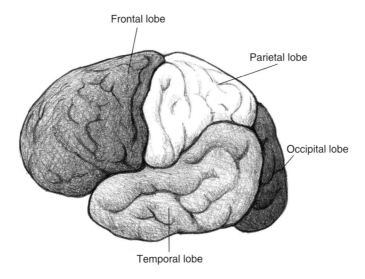

Frontal lobe

Parietal lobe

Occipital lobe

Temporal lobe

Figure 8.3 **Sketch of a human brain, showing the lobes**

activity, we can only track the evolution of a few of these in fossils because our knowledge of the brain surface of fossil hominids is limited by various circumstances.

First, we have no fossil brain, only casts (natural or artificial) of the cranial cavity, and although the internal walls of the cranium reproduce the general morphology of the brain surface they do not do so with sufficient detail to identify the exact limits of most areas of the brain with any precision. This is because the brain is covered by a triple layer of membrane, the meninges.

Moreover, the larger the brain of an organism, the more faint and indistinct the impressions of the brain on the inner surface of the cranium. Some regions of the brain surface also leave more detailed impressions than others. Thus, for example, the impressions of the occipital lobe are lighter than those of the frontal lobe. In short, studies of the paleoneurology of the first hominids are limited to a few specific aspects of brain morphology.

The two world authorities in this area, Ralph Holloway and Dean Falk, hold different opinions on the brain of australopithecines and *Paranthropus*. Holloway believes that these hominids already show evidence of reorganization of the brain compared with apes; these are reflected, for example, in a degree of development of *Broca's area* (an area related to the production of speech, which we shall look at in the chapter dedicated to language). However, Falk takes the view that the brain surface of *Paranthropus* and australopithecines is clearly similar to that of apes and that there is no evidence of reorganization of the brain.

Despite their differences on this matter, Holloway and Falk agree that there are clear indications of cerebral restructuring, similar to that of modern humans, in the fossils of the first representatives of *Homo*. These indications can be summed up in two major processes: the appearance of marked asymmetry between the cerebral hemispheres, and a more complex morphology of the frontal lobe.

The human brain is clearly asymmetrical, and although apes, and anthropoids in general, show cerebral asymmetries, these are not of the same kind or as extensive as those typical in humans. The asymmetries characteristic of the human brain are not directly related to the functional specialization of the cerebral hemispheres observed in humans. This phenomenon is known as *brain lateralization*, and seems to be confined to our species. To generalize, the left hemisphere specializes in functions related to language, precision movements of the right hand, the capacity for analysis, and the perception of temporal sequence. The right hemisphere is devoted more to tasks such as the skill of the left hand, the capacity for overall understanding of processes, recognition of faces, spatial vision, musical skills, the control of the tone of the voice, and the expression of emotions and recognition of them in others.

In addition to other functions (such as control of primary motor functions), the frontal lobe is responsible for a series of mental capacities which are exclusive to or most fully developed in humans. These include the capacity to determine the sequence of movements of the phonetic apparatus which make up speech, control of the emotions, the ability to concentrate on a task, planning, anticipation of events, maintaining an idea in the mind for a long time, and control of the use of the memory to integrate previous experiences and learning when taking decisions. Metaphorically speaking, we could say that the frontal lobe is the "conductor" of the orchestra of our brain.

Over the course of human evolution the frontal lobe has expanded a great deal, both in absolute terms and in relation to the rest of the brain. But recent investigations by a group of scientists led by Katerina Semendeferi indicate that the volume of our frontal lobe is what would be expected in a primate with a brain the size of ours. Given that the frontal lobe appears to be the most "human" region of our brain, perhaps the increase in brain size is merely the result of the selective advantage that having an increasingly more developed frontal lobe conferred on our ancestors. In other words, it is reasonable to suppose that the large size of the brain in our species is not the cause, but the result of our large frontal lobes. If this is the case, development of the frontal lobe would have been the driving force in our encephalization.

What we can at least say with certainty is that the surface of the frontal lobe has become increasingly complicated, owing to an increase in the number of sulci which cannot be explained purely in terms of the greater size of the human brain.

These characteristics – the cerebral asymmetry linked to functional lateralization, and greater structural complexity of the frontal lobe – are clearly marked in the endocranial casts of *Homo habilis/Homo rudolfensis* and *Homo ergaster*. So even in the first humans, in addition to an increase in brain size, we find clear indications of a brain structure similar to our own.

To complete our examination of the brain of the first hominids, we might speculate as to the nature of the vital advantage the new type of brain gave our ancestors, favoring their selection. There are two hypotheses on this subject.

The first, put forward by Robin Dunbar and Leslie Aiello, links the *increase in the neocortex* (i.e., the cerebrum) to the improvement in social skills within the group. According to these authors, the increase in size and the reorganization of the human brain are linked to the development of "social intelligence." We shall return to this hypothesis when we look at the social biology of the first hominids.

Falk, on the other hand, relates the modifications in the brain of the first humans to a very specific capacity, that of *language*. She argues that given that one of the crucial centers of human speech is located in the frontal lobe, and that the production and decoding of language are functions which are clearly lateralized in the human brain, both lateralization and the increased structural complexity of the frontal lobe are related to the development of the linguistic capacities of our first ancestors. It is immediately evident that these two hypotheses are complementary, since the primary social skill is perhaps the capacity to communicate efficiently with others.

However, the linguistic capacities of the first humans are the subject of major controversy. We shall return to this subject in the chapter on language.

The Size of the Intellect

During the last quarter of the 19th century it was fashionable in some scientific circles to link human intelligence with brain size. During that time, a series of measurements of brain size was made on a number of skulls and cadavers, with the aim of identifying the variations in intelligence in humans. These studies were principally the work of Paul Broca (1824–80), Professor of Surgery at the Faculty of Medicine in Paris, and his followers. One of the most significant results of these studies was the finding that on average, women have smaller brains than men. This fact was used to assert that, because of their smaller brain mass, women are intellectually inferior to men.

Today, at the end of the second millennium, this story might appear outdated to many of our readers. However, the argument that women's smaller brain mass indicates reduced intelligence still persists to some extent today.

As we have seen in this chapter, it is not overall brain size which determines the size of the intellect, but the proportion of brain size to body size, or the degree of encephalization. Women's brains are, on average, smaller than those of men because women's bodies are also, on average, smaller. Both women and men have brains which correspond to their respective body size; in other words, the two sexes are equally encephalized.

9

Teeth, Guts, Hands, Brain

Compare, at entirely similar ages, a man who, in order to free himself for studies and habitual intellectual work, has acquired the habit of eating very little with another who habitually takes plenty of exercise, frequently goes out of the house, and eats well. The stomach of the first will have very little capacity and will be filled by a very small quantity of nourishment, while the stomach of the second will have preserved and even increased its capacity.

Jean Baptiste de Lamarck, *Zoological Philosophy*

Types of Diet

There is ongoing debate in our society as to whether we are "naturally" vegetarians or carnivores. In fact many campaigning vegetarians include milk and dairy products, which are derived from animals, as well as eggs, in their diet. Given that, on the other side, there is virtually no one who is exclusively carnivorous by conviction, the debate is generally between ovo-lacto-vegetarians and omnivores, who in addition to eating vegetables, eggs, and dairy products also eat meat and fish. Before going further, we must make it clear that both diets have been scientifically proven to be healthy, provided that they are balanced. However, most specialists would not say the same of a strict vegan diet.

Nevertheless, there are those who take the view that as apes, our diet should consist only of vegetables, ignoring the fact that many primates eat greater or lesser quantities of invertebrates, particularly insects, and small vertebrates. But since primates vary so widely, perhaps we should restrict ourselves to looking at our closest relatives. It is true that the diet of the great apes consists almost exclusively of vegetables, mainly ripe fruit, leaves, stems, and tender shoots. Orangutans and chimpanzees eat more fruit, or are more frugivorous, than gorillas, which eat more leaves, shoots, and stems, and are thus more folivorous.

Everything has its advantages and disadvantages; fruits are considerably more nutritious because they contain more simple carbohydrates like glucose, fructose, and sucrose. Leaves and tender stems are probably more abundantly available, but contain more fiber, which is carbohydrate that we cannot assimilate, such as cellulose (although fiber is also important for good digestion). However, although no one would classify chimpanzees as predators, they have been seen hunting monkeys and young antelope and other small mammals. They also eat insects such as caterpillars, ants, and termites. To sum up, while they may be basically vegetarian, they seem to like animal proteins.

In the case of our species we cannot rely on "field" studies to resolve the problem of man's natural diet in his environment. Since agriculture and animal husbandry were "invented" during the Neolithic period, 10,000 years ago, animal and plant foods have been eaten in varying proportions depending on the culture in question. Nevertheless, the majority of humanity has lived, and continues to live, principally on the hard grains of cereals – wheat, rice, corn, barley, oats, rye, millet, and so on. Another important element in human sustenance is the dry seeds of legumes (beans, peas, lentils, etc.). All of these grains and seeds (cereals and legumes) are rich in long-chain carbohydrates (particularly starch), which are made up of many molecules. Legumes contain more protein (around 20 percent of their mass) than cereals (about 10 percent), but in both cases these are low-quality proteins in the sense that they are incomplete. Let us look at what this means.

Proteins are long chains of molecules known as amino acids. Of the 20 types of amino acids which exist in proteins, humans can only synthesize 11. We have to obtain the others, known as the essential amino acids, from food. Five of these amino acids are less abundant in foodstuffs, and some are not found at all in some cereals and legumes, although all can be obtained from an appropriate combination of various types of these plant products.

It is worth pointing out that cereals and legumes are not eaten as they are found in nature: they need to be prepared. The minimum required to prepare them is tools for grinding, to produce flour, and fire, water, and a ceramic vessel to cook and soften them. Thus to some extent they are not "natural" foods.

The remainder of the vegetable element in the human diet consists of tubers, like potatoes, which are rich in starch, and vegetables. Vegetables comprise different parts of plants, from leaves (like lettuce and spinach), to bulbs (such as onion, garlic, and leeks), roots (carrots or beet), flowers (cauliflower and artichokes), stems (like celery), and fruit (such as tomatoes and pimentos). Vegetables contain a lot of water and few calories. Finally, we have fruit. Although some vegetables are eaten with little preparation, only fruit is eaten as found in nature, without adding anything or cooking it.

Nuts, which are fruits with shells (such as almonds, hazelnuts, and walnuts), and which are very nutritious because of their high fat content, but fairly well protected against our fragile teeth, merit separate mention. Incidentally, while a hazelnut is a nut in the botanical sense, almonds and walnuts are in fact fruits with an internal stone which contains the nutrients (technically they are drupes, or stone fruit, like cherries and olives, except that in the case of the latter we eat the fleshy envelope).

The total amount of protein we require is fairly small: even highly trained athletes need no more than 1.5 g per kilogram of body weight, and for most of us the total required is less than 100 g. Thus we do not need to eat a large quantity of animal products, and the amount consumed in First World countries is clearly excessive.

So we see that there are major differences between the dietary habits of apes and our own, at least since Neolithic times. Comparing apes' diet with human diet, we could say that apes eat vegetables and fruit, while we (as a species) live on the tough grains of cereals – a form of grass – and the seeds, also dry, of legumes.

There are some human groups that have been identified and studied within recorded history which do not engage in agriculture or animal husbandry. It is worth noting that they hunt, fish, and gather very varied animals and plants. What they eat varies in relation to a number of factors, such as the resources available in the environment and the time of year, but no cases of groups living either exclusively on what they hunt or exclusively on plant foods have been found. The only exception is perhaps the Inuit, whose economy until recently relied almost entirely on hunting and fishing; however, this is an exceptional case related to the special characteristics of their environment, which is covered in ice and snow for much of the year and devoid of any vegetation.

How do we approach this question from the paleontological point of view? By studying the only part of the body directly related to diet which fossilizes – the teeth, and together with them, the chewing apparatus.

Carnivorous and Herbivorous Mammals

Carnivorous mammals have adapted their teeth to the functions they need them to perform. Carnivores vary widely in form, but all of them have certain teeth specialized in cutting meat, known as carnassial teeth (these are the last upper premolar and the first lower molar). Carnassial teeth have a very sharp edge which allows them literally to slice up the flesh of dead animals (Figure 9.1).

Proteins and animal fats are rich in energy, having a high calorific value, and they are also easy to digest. This means that this type of food does not need to be prepared in the mouth before it is swallowed, so chewing is brief, its purpose being only to reduce the flesh of the dead animal to pieces of a size that can pass through the esophagus. The carnassial teeth operate like shears, or scissors, the upper one passing to the outside of the lower when the mouth is closed. This scissor-like movement causes the carnassial teeth to sharpen themselves on one another, and as they wear against one another they become razor-sharp. Unlike our kitchen knives, this wear improves the cutting edge of the carnassial teeth and increases their efficiency, rather than dulling or chipping them.

Carnivores also need tools to kill their prey, and the canines, often known as eye-teeth, are highly developed in order to perform this function. The terrestrial carnivores with the most highly developed canines are the Felidae, the family which includes cats, lions, tigers, and so on. In these animals the incisors are small and the canines very large; the carnassial teeth are highly developed and are the hindmost teeth in the jaw (although the vestige of a molar remains behind them).

Hyenas have a similar dentition to that of the big cats: in fact their diet is similar to that of the Felidae, although they do not always kill their prey. (It has been established, however, that the larger spotted hyena is a powerful group hunter.) In addition to the range of teeth seen in Felidae, hyenas also have highly developed conical teeth (the penultimate premolar, both upper and lower), which they use as hammers to crush the bones of large herbivores. When the predators and the majority of carrion eaters have finished eating, the tougher bones still remain: these enclose a very fatty, and therefore energy-rich, material, the marrow. Hyenas can gain access to this resource using their specialized premolars. At a certain point, some other animals also began to compete for the marrow contained in the strong-walled bones of the large herbivores; these were primates – humans, who, lacking the appropriate teeth, used stone tools to split the bones.

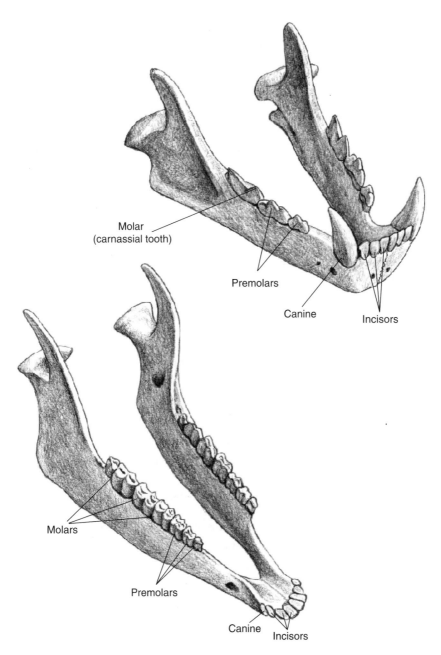

Molar
(carnassial tooth)

Premolars

Canine

Incisors

Molars

Premolars

Canine
Incisors

Figure 9.1 Jawbones of (*top*) a lion; (*below*) a sheep

But not all land-based carnivores have such a specialized diet, and therefore dentition, as the cats. The Mustelidae (the family which includes badgers, otters, martens, mink, and weasels), the Canidae (the group of the foxes, wolves, coyotes, African wild dogs, etc.), the Ursidae (bears), and the Viverridae (such as genet and mongoose) have molars behind the carnassial teeth with chewing surfaces instead of cutting edges. While the front part of the lower carnassial teeth has a cutting edge, the back does not. This indicates that their diet also includes vegetable matter which is more difficult to absorb, and requires prior chewing, insalivation, and predigestion in the mouth. The famous cave bear, a gigantic animal bigger than any modern bear, which lived at the same time as *Neanderthal* and *Cro-Magnon man*, was basically vegetarian, as its large molars indicate.

It goes without saying that the large herbivores also modified their teeth to the function they had to perform. The great grasslands – the steppes, prairies, and pampas – provide enormous quantities of plant food which is, however, of low nutritional value because the plants which make it up are rich in fibers like cellulose. In order to absorb the food, a high level of predigestion in the mouth is required, where the fibers are reduced to a pulp which is then processed in the digestive tract. This processing requires the assistance of symbiotic microorganisms (basically bacteria) which live in the intestines of herbivores and can perform the miracle of decomposing cellulose into carbohydrates which can be absorbed by the animal.

The teeth of these mammals have a ridged crown which increases the efficiency of their chewing surfaces. The work of maceration and reducing the vegetable matter to a pulp is speeded up if the surfaces rubbing against one another are ridged, like millstones in traditional grain mills (Figure 9.1).

The premolars and molars of herbivores have to withstand intensive wear, since the fibrous stalks of plants like grasses also contain mineral particles which make them harder and more abrasive. They therefore have very high crowns, so that they last longer, and very wide chewing surfaces; they have no cutting edge for slicing flesh.

The Teeth of Primates

What kind of teeth do our closest relatives, the great apes, have? And are ours similar to those of herbivores, or carnivores, or like those of the great apes? The latter have very large incisors and canines, particularly the males.

This development of the canines is common to all anthropoids, particularly in males, and it does not indicate an adaptation to a carnivorous diet. The canines are used for fighting, both within the species and against predators. Baboons, geladas, and mandrills are well known for the males' enormous canines, which leads even successful predators on anthropoids, such as leopards or even lions, to treat them with respect. However, they cannot be considered in any way carnivorous, although they do on occasion catch insects, small vertebrates, and baby mammals.

The premolars of the Old World monkeys and the apes have two cusps (like those of humans), but the lower anterior premolar is long and narrow, and the outer cusp is so much higher than the inner that it can be considered to have only one cusp. The posterior edge of the upper canine slides against the anterior edge of the first lower premolar, creating a scissor-like action. This type of anterior premolar with a cutting action is known as *sectorial*. There is also usually a space between the upper lateral incisor and the upper canine, into which the lower canine inserts when the mouth is closed; this is known as a *diastema*.

The molars are the most interesting teeth in terms of inferring type of diet. There is a fundamental difference in the crowns of the molars between the Old World monkeys and the apes. In Old World monkeys each molar has two transverse ridges. These monkeys have a wide range of diets, but as we have seen, the ridged form of the teeth is typical of mammals which eat plant products requiring a degree of predigestion in the mouth. In particular, the group of the colobus monkeys and langurs, which eat large amounts of leaves, have teeth with higher cusps and sharper ridges.

In apes the cusps, four on the upper molars and five on the lower molars, form isolated projections separated by dips (without the transverse ridges seen in the Old World monkeys). To see what this actually looks like, you can look at your own teeth in a mirror (assuming that the relief of the crowns has not been altered by fillings) (Figure 2.1). This shape is not specialized in any particular way, making it fairly well adapted to the general type of diet of apes, which includes fruit and also the less tender parts of plants. Gorillas especially eat the tougher parts of plants, since they have become too large to climb trees looking for fruit: in line with their fiber-rich diet, gorillas have higher cusps on their molars.

The great apes in general, and particularly chimpanzees, have very large upper central incisors, with a straight cutting edge so that they function like chisels. The upper lateral incisors are smaller, the lower incisors even smaller, but all have a straight cutting edge and the same chisel function. It is very difficult to eat an apple using only your molars and premolars.

The front teeth (incisors and canines) work to slice up fruits of a certain size so that they can then be chewed by the back teeth (premolars and molars).

We modern humans have modified the dentition of our ancestors considerably. To begin with, our incisors are not as large as those of the apes. We have no diastema, and the tips of the canines project hardly at all beyond the chewing surfaces of the other teeth; moreover, with wear the point of the canines very quickly becomes level with the surface of the adjacent teeth. In humans and apes the incisors are spatulate, or fairly flat, widening from the base of the crown (or neck) to the cutting edge. In apes the canines are conical, but in humans their shape is closer to that of the incisors, spatulate, and becoming wider until just below the tip of the tooth (which is in any case fairly rounded).

In humans the first lower premolar has two cusps like the others, and it does not slide against the posterior edge of the upper canine. In apes the incisors are far in front of the canines, which are aligned with the premolars and molars, so that the dental arches have parallel sides and are U-shaped. In humans the dental arches are parabolic or elliptical.

The Teeth of the First Hominids

Of all the fossil hominids, the one which is most similar to living apes is *Ardipithecus ramidus*. In this species the canines project considerably, although not as much as in apes, and the first lower premolar is sectorial. Other typical features of apes which have become lost over the course of human evolution are found in this hominid, including the morphology of the dental arch, which is U-shaped with parallel sides. No palates have been preserved, but judging from the size of the canines we can imagine that there was a well-developed upper diastema.

In *Ardipithecus ramidus* the first milk molar has a very prominent principal cusp, as in apes; this characteristic had never before been seen in a hominid (all the others have several cusps of similar size). This trait is so significant that a small fragment of jawbone showing a conical first milk molar appeared on the cover of the historic issue of *Nature* in which, on September 22, 1994, the discovery of *Ardipithecus ramidus* was published.

We have already noted that the enamel is also very thin in this species. We have to conclude that the diet of *Ardipithecus ramidus* differed little from that of modern chimpanzees. However, there is one trait which merits further attention. Although the canines project much further than in modern humans, they do not project as much as in apes, nor are they the same shape. As we have mentioned, in humans the canines are spatulate or incisiform (having the shape of the incisors), and this condition appears already incipient in *Ardipithecus ramidus*. Moreover, the anterior edge of the first lower premolar is not sharpened through wear against the upper canine (as it is in apes). In fact, this change in the function of the upper canine/first lower premolar pair, small though it is, is one of the main arguments for classifying *Ardipithecus ramidus* as a hominid. This combination of traits prompted Tim White and his colleagues to attribute a crucial role in human evolution to this species, as ancestor of all the subsequent hominids, and thus our first known ancestor. Given its antiquity (4.4 million years), it is unlikely that much more primitive forms of hominid will be found.

Although when the first fossils of *Australopithecus afarensis* were discovered they seemed the height of primitivism (as had *Australopithecus africanus*, long before), the archaic traits of *Ardipithecus ramidus* give an idea of how many things had changed in *Austalopithecus afarensis*. Some of the fossils of the latter still show first lower premolars with a very prominent cusp (while others already have two cusps), the diastemas in the upper dental arches are small or nonexistent, and the canines project less and are more incisiform than in *Ardipithecus ramidus*. The most important feature is that in *Australopithecus afarensis* the molars have become larger, and particularly wider, than in *Ardipithecus ramidus*. This, combined with the increase in thickness of the tooth enamel, indicates that a major ecological change had taken place, resulting in a diet which is still plant-based, but now contains a large proportion of tough, abrasive foods. There is no trait in the dental morphology which indicates significant consumption of flesh. Our first ancestors, both those who lived in the humid forest like *Ardipithecus ramidus* and those who had begun to exploit the resources of dry woods and clearings like *Australopithecus afarensis*, were not hunters but vegetarians. To put it more dramatically, the first hominids did not come down from the trees to become "murdering apes," nor was it their taste for meat that prompted them to abandon the forest.

The fossil record for *Australopithecus anamensis* is still scarce, but we have some information on the masticatory apparatus of these fossils which are intermediate between *Ardipithecus ramidus* and *Australopithecus afarensis*. The Kanapoi maxilla and mandible, together with the teeth that have been preserved, show very primitive characteristics, but it appears that the enamel had already become thicker and the size of the molars increased to a certain extent.

The traits which distinguish *Australopithecus afarensis* from *Ardipithecus ramidus* are still more marked in *Australopithecus africanus*; for example, in the latter all the first lower premolars have two cusps.

Size of the Molars and Shape of the Hand

One characteristic of the masticatory apparatus that we have so far mentioned only in passing is closely related to the type of diet: this is the size of the chewing surface. This variable also depends on the size of the individual, so we need to separate the two factors (body size and type of diet). If a species shows a large difference in size between the sexes, the males will have larger teeth than the females – not only the canines, which are not directly related to diet, but also the premolars and molars, which are. This is due simply to the fact that having a larger body to maintain means needing to process more food, or to increase the amount of chewing done. This is why gorillas have bigger teeth than chimpanzees. But gorillas' teeth also have more work to do than those of chimpanzees because gorillas include in their diet plant products which are more fibrous and less nutritious than the fruits which chimpanzees eat. Scientists therefore have to work out how to establish how much of the difference in tooth size between the two species is due to difference in body size, and how much is due to the type of diet.

Clearly this is a similar problem to that of relative brain size which we have already discussed, and it is resolved in the same way. We can construct an *index of megadonty*, or tooth size ratio, between the actual size of the molars in a given fossil species (this value forms the numerator), and that which should correspond to a primate of the same size (the denominator). The latter, hypothetical value is calculated using an equation which relates the size of the molars to body size for a set of modern species. We can also approach the problem more directly: an average female *Australopithecus afarensis*, which would weigh a little less than an average female chimpanzee, should have slightly smaller molars than the female chimpanzee if her diet was the same. However, those of the fossil species are larger, indicating a different diet and more laborious chewing in *Australopithecus afarensis*. The molars of *Ardipithecus ramidus*, on the other hand, are smaller than those of *Australopithecus afarensis*, and comparable to those of chimpanzees, as their diet must also have been.

Using the method described above, Henry McHenry has calculated an index of megadonty (he measures the chewing surface for the set of teeth formed by the last premolar and the first two molars). The value of the index for *Australopithecus afarensis* is almost double that for chimpanzees, indicating that diet had already changed to include tougher plants. The relative size of the chewing surface increases somewhat in *Australopithecus africanus* and continues to increase in *Paranthropus robustus* and *Paranthropus boisei* (to three times that of chimpanzees). However, in *Homo habilis* the value of the index returns to the level of *Australopithecus afarensis*, and in *Homo ergaster* it approaches that of the chimpanzee, very similar to our own.

Paranthropus are the hominids with by far the largest chewing surface, both in absolute terms and in relation to body mass. And not only the molars, but also the premolars are larger, particularly the posterior premolar, which has become "molariform," increasing the number of cusps and acquiring the form of a molar. The incisors and canines, on the other hand, have become very small, creating an imbalance between the very reduced anterior teeth (incisors and canines) and the enlarged posterior teeth (premolars and molars).

Large canines which project beyond the level of the adjacent teeth and have to be slotted into diastemas make it difficult to move the front part of the jaw from side to side. This type of lateral movement combines with the vertical movements of opening and closing the mouth to produce rotation. Rotational movements are necessary to grind tough plant foods. The great apes do not have this problem because their diet does not require so much grinding. In baboons the face is so large that some rotation of the posterior part of the jaw is possible, although in males the anterior part is blocked by the hugely developed canines. In *Australopithecus*, *Paranthropus*, and *Homo*, the wear on the premolars and molars is very intensive; as soon as these teeth appear their chewing surfaces characteristically become flat and with virtually no relief, the enamel disappearing rapidly and wearing closer and closer to the dentin, the tissue beneath the enamel.

What kind of diet did *Paranthropus* have? No doubt similar to that of modern baboons, based on hard grass seeds, fresh or dry legumes, and nuts (Figure 9.2). It would also have included fruits like blackberries, berries, and stone fruit, which were easier to chew, and the underground storage organs of plants like bulbs, tubers, fleshy rhizomes, and tuberous and swelling roots (baboons eat practically everything, including small animals).

We have already noted that grass seeds and legumes are rich in carbohydrates (which they store principally in the form of starch), and are thus very nutritious, although low in protein (legume proteins, moreover, have to be cooked in order to be absorbed). As processing seeds in the mouth is very laborious, *Paranthropus* developed their own personal mill, through

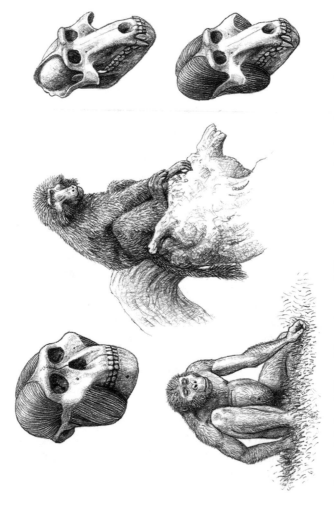

Figure 9.2 *Paranthropus boisei* and *Theropithecus brumpti.* These two primates developed somewhat similar specializations in their chewing apparatus, owing to a similar diet, containing a high proportion of tough plant matter

the expansion of the premolars and molars and of the muscles which moved them and the bones in which the muscles were anchored. This powerful chewing apparatus would also have been useful in cracking nuts, rich in vegetable oils. The underground storage organs of some plants provide carbohydrates and some proteins, although the soil which they would inevitably have ingested with them would have contributed to wear on the teeth. Baboons, and particularly their relatives the geladas which live on the high, virtually treeless plains of Ethiopia, can digest grasses and grass stems, but this is a unique specialization which cannot be assumed without evidence to have existed in hominids.

Some authors believe that *Paranthropus*, and also the australopithecines, used sticks to dig bulbs or tubers, for example, out of the earth. It is also possible that both genera included animal products in their diet, as do baboons and chimpanzees when they can. However, the theory maintained by Dart and Le Gros Clark, that the first hominids were predators on a large scale, using tools to kill their prey, is now completely discredited. In a magnificent example of taphonomic analysis, Bob Brain demonstrated that the hominids whose bodies accumulated in the caves of South Africa were not the fierce occupants of the caves, but rather prey brought there by the real hunters, leopards. The baboons, gazelles, and other herbivores whose remains are found together with the hominids would have met the same fate. Lee Berger and Ron Clarke have recently postulated that even the Taung Child was the victim of a predator, although this time a winged one – an eagle.

One interesting aspect of baboons' diet is that it requires a certain capacity for manipulation of small objects, such as seeds, with the hand. There is no published data on what *Ardipithecus ramidus*' hands were like, but it is possible that in the first hominids (this species or an earlier one) the morphology of the hand was similar to that of apes, with a short thumb and long palm and fingers. Remember that this is an adaptation for hanging from trees by forming a hook with the hand.

In *Australopithecus afarensis* the hand is already almost the same as ours, although the phalanges are more curved, somewhat like those of chimpanzees; moreover, the thumb is proportionately shorter and the rest of the hand slightly longer than in our own species. Hand bones of a modern type have been found in the Swartkrans deposit. Most of the Swartkrans fossils are from *Paranthropus robustus*, but as there are also some from *Homo*, we cannot be sure to whom these hand bones belong. However, we can venture the hypothesis that, at least since *Australopithecus afarensis*, hominids had modified their hands so as to have full capacity for manipulation of small objects, as a result of their new diet.

Our vision of the first hominids has changed greatly with time. Originally they were envisaged as fierce predators of the savanna, and then as inoffensive vegetarians, also living in the savanna. *Ardipithecus ramidus* is now believed to have lived in the humid forests, and all species of *Australopithecus* are thought to have been vegetarians living in a mixed environment, whether this is called dry forest and clearing or densely wooded savanna. Two types of hominids which were genuinely specialized in open environments, *Paranthropus* and the first *Homo*, seem only to have appeared 2.5 million years ago.

Guts and Brain

In 1891 Sir Arthur Keith made an observation which passed unnoticed at the time. He had noted that in primates there was an inverse relationship between brain size and stomach size. Surprisingly, the larger the stomach, the smaller the brain – in other words, a primate cannot have a large digestive system at the same time as a large brain. This fact urgently required explanation. However, we had to wait for this explanation for over a century until, in 1995, Leslie Aiello and Peter Wheeler put forward a hypothesis which is extremely important for studies of human evolution.

Aiello and Wheeler point out that, since the brain is one of the most costly organs in terms of metabolism (the economy of the body), an increase in the volume of the brain would only be possible if it was balanced by a reduction in the size of some other organ with a similar energy consumption. In proportion to their mass, the organs with the highest energy consumption in the human body are the heart, the kidneys, the brain, and the ensemble formed by the digestive tract and the liver. The brain accounts for 16 percent of the organism's basal metabolic rate (the energy consumption, per unit time, required to maintain the vital functions of an individual at rest); the digestive tract represents a similar proportion, at 15 percent. We have already seen, in the section on encephalization, that human beings have a brain substantially bigger than that which corresponds to a hypothetical nonhuman primate of our size; it turns out that our digestive tract is smaller than that which would be expected by almost exactly the same proportion.

Aiello and Wheeler conclude that the enlargement of the brain which occurred in *Homo* was only possible with a shortening of the digestive tract. The length of the digestive tract depends on the kind of food it has to process: in carnivores it is always shorter than in herbivores because meat is easy to digest. Herbivores, on the other hand, need long digestive tracts in order to metabolize the plants they eat, particularly if these are rich in cellulose.

We have already noted that about 2.5 million years ago two types of hominid settled in open environments. This is a point in time which many authors consider important from the point of view of climate change, because a general cooling of the planet occurs – reflected in Africa by the definitive expansion, at the cost of the enclosed forest environments, of the great grasslands and the savannas (albeit still more or less wooded). One of these two types of hominid is *Paranthropus*, which adapts its chewing apparatus to eat the tough but nutritious plants of the savanna, as baboons do today.

However, the brain of *Paranthropus* does not increase in size as much as that of *Homo*. Bearing in mind that this increase in brain size implies an increased energy consumption, only two solutions remain. One is increasing the basal metabolic rate of the organism as a whole (the total energy consumption). This is not what happened, since humans have the rate which corresponds to a mammal of our size. The other solution is to reduce the energy consumption of another organ in order to balance the body's energy economy. The question is, which organ will have to reduce in size? Not the heart, nor the kidneys, nor the liver, which are vital parts. However, the digestive tract can become smaller if diet is improved, by increasing the proportion of high-quality nutrients, those which can be easily digested and have high calorific value. What are these high-quality products which do not form part of the diet of *Paranthropus*? The only answer can be animal fats and proteins. The first humans must have shifted to incorporating a larger proportion of meat than any other primate in their diet; they would have got this first as carrion-eaters, and then increasingly as hunters.

For the first time in the history of mammals, this change in diet was not reflected in a change in dental morphology. We do not find humans with teeth that work like hammers to crush bones, nor with teeth that act like knives to cut meat, because the tools required for splitting bones and cutting skin and meat are outside the body: they consist of stones and the edges of stones worked by humans.

Thus, the enlargement of the brain in *Homo* could only have occurred with a change in diet; this in its turn is reflected in the reduction in size of the digestive tract and a correlative reduction in the chewing apparatus. Aiello and Wheeler insist that this does not mean that change in diet automatically leads to an increase in brain size; they maintain only that we had to become carnivores in order to become intelligent (although this is a vicious circle, since locating high-quality food requires greater mental capacity).

The last *Paranthropus* disappeared in Africa, where they emerged, about one million years ago. Perhaps their ecological niche was continually being reduced through competition with baboons and geladas and with humans, our ancestors. The curious aspect of this is that, since the development of agriculture, most of humanity has sustained itself largely on plant products which, although cultivated, are fairly similar in composition to those eaten by *Paranthropus*. The difference is that we do not grind the hard seeds of cereals and legumes with our teeth, or crack nuts with them; since Neolithic times we have been cooking the seeds or making them into flour using artificial grinders. We learned to crack nuts with a stone long before this.

10

Development

*Yes. One goes on. And the time, too, goes on – till one perceives ahead a
shadow-line warning one that the region of early youth, too, must be
left behind.*

Joseph Conrad, *The Shadow Line*

The Rhythm of the Molars

Broadly speaking, the four main stages of life in catarrhine primates
(which include ourselves) are marked by the eruption of the three molars
of our adult dentition. The emergence of the first molar marks the end of
what we might call early childhood, a period of strong dependence on the
mother, from whom the child is rarely separated, and which basically
covers the period of nursing (only in humans does it include several
more years of development).

The appearance of the second adult molar coincides with the end of the
second period of childhood, and the beginning of the important changes of
puberty, in males an increase in the level of testosterone in the blood is
detected shortly afterward.

The emergence of the third molar, known as the wisdom tooth, corresponds to the completion of development and the beginning of adult
life, although the final fusion (knitting) of all the bones occurs somewhat
later.

While the three major periods of development are essentially the same
in all primates, their duration varies. In humans development continues

over about twenty years, almost double that of the great apes (chimpanzees, gorillas, and orangutans). On average chimpanzees get their first adult molar when they are a little over three years old; in our species it erupts at the age of six. In chimpanzees the second molar emerges at the age of six and a half, while in humans it appears at around eleven years; finally, the third molar appears in chimpanzees at around age eleven, and in ourselves at around eighteen (although among modern humans the four wisdom teeth sometimes do not appear at all, and when they do it can be well into adulthood). In chimpanzees the first period of estrus, with its characteristic swelling of the ano-genital region, occurs between the ages of nine and almost fourteen, depending on the nutritional condition of the female (in humans too the first menstrual period, or *menarche*, occurs earlier, on average, in girls who are well fed and healthy). In chimpanzees as in humans, this is usually followed by a period of infertility before the first conception. In chimpanzees the first birth occurs on average at the age of fourteen. A chimpanzee can live to the age of forty or more.

Despite this difference in length of the periods of development between apes and ourselves, the relatively close correspondence between the stages of life and the appearance of the molars helps us to establish the state of development of a fossil hominid: although we do not know its "chronological age" (i.e., how old it was when it died), we do know its "physiological age" (whether it was an infant, an older child, an adolescent, or an adult).

Thus the famous Taung Child was emerging from early childhood when the eagle snatched him, because the first molar was erupting precisely at that moment. So how old would he have been? If he died between the ages of three and four this would mean that the rhythm of development of *Australopithecus africanus* was similar to that of chimpanzees, while if he died at six, it would be slower, like our own. When Turkana Boy died the second molars had already erupted. Was he around eleven years old, like a modern child in terms of the development of his skeleton and teeth, or was he around seven years old, like a chimpanzee with the same "physiological age"? How can we find out? In order to tackle this question, we shall start at the beginning of existence – birth.

Birth and the Newborn

Shortly before birth the fetus shifts into an inverted position, with the head in the upper part of the mother's pelvis (formed by the iliac crests and known as the *greater pelvis* or *false pelvis*). During birth, the full-term fetus has to pass through the lower part of the pelvis (known as the *lesser* or *true pelvis*) via a bony tube called the *pelvic cavity* or *birth canal*. Among chimpanzees, gorillas, and orangutans birth is easy and rapid, because the birth canal is large in relation to the head of the fetus. In these apes the entry to the birth canal (known as the pelvic inlet) is oval, with its major diameter running from front to back (a sagittal or anteroposterior orientation, in anatomical terms), and its minor diameter oriented transversely (from side to side). As we have seen, the modification in pelvic architecture required to make it possible to walk upright meant that the joints between the coxal bone and the spine, and the coxal bone and the femur, drew closer together. As a result, the sagittal diameter of the birth canal reduced, and this is why complications can occur in the human birth process.

In human females the entry to the birth canal is not oval but round (Figure 10.1). The major diameter of the pelvic inlet is not sagittal: it is sometimes transverse but more often neither sagittal nor transverse, but on the two oblique or diagonal diameters. Moreover, the head of the full-term fetus is elongated, so that its major diameter is anteroposterior (from the forehead to the nape of the neck). The position of the fetus' head adapts to the greater diameter of the pelvic inlet, generally one of the oblique diameters. As the greatest dimension of the outlet to the birth canal is always sagittal (in humans as well as in other primates), the skull and shoulders of the human fetus have to enter the birth canal with one orientation (transverse or oblique) and leave with another (sagittal). Thus there is a rotation within the birth canal both of the head and, afterwards, of the shoulders (Figure 10.2).

To complicate matters further, in humans the vagina is angled forward, forming a right angle with the uterus, so that as it passes through the birth canal the fetus moves not in a straight line, but around a pronounced curve which ends immediately below the pubic bone, where the baby's head emerges. In order to accommodate this curved trajectory, the fetus' spinal column arches, the head flexes strongly toward the back and the crown of the head faces forward as it is born, so the baby is facing backward (in the opposite direction to the mother). In other primates the vagina lies at the same angle as the uterus, with which it is aligned, and the full-term fetus follows a straight trajectory toward the back as it is born; furthermore the face is forward (in the same direction as the mother). To sum up, birth in humans is ventral, while in other primates it is dorsal.

Karen Rosenberg and Wenda Trevathan have pointed out that the mother ape can help her baby to be born, guiding it through birth with

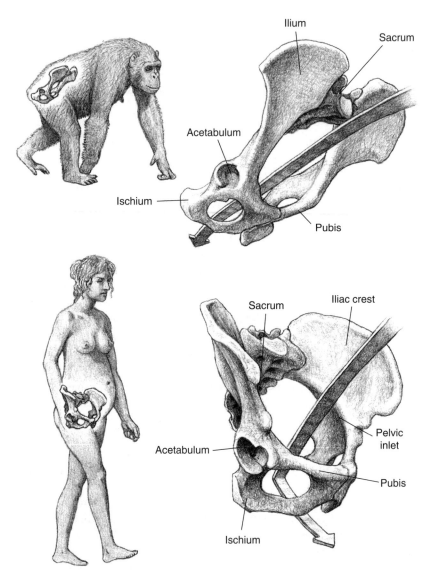

Figure 10.1   Morphology of the pelvis in a chimpanzee and a human female.
The arrow indicates the direction taken by the full-term fetus during birth

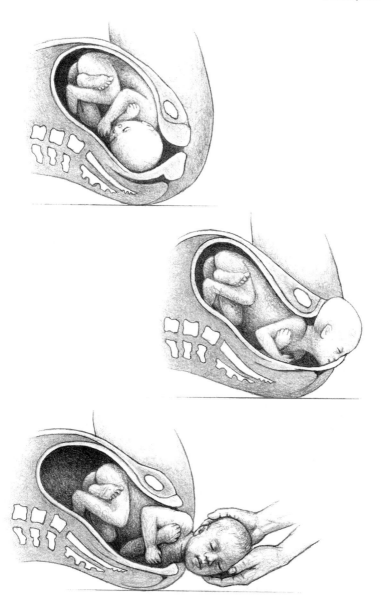

Figure 10.2 Flexion and torsion of the fetus during birth. To begin with (*top*), the head lies in a transverse or diagonal position and the shoulders in a sagittal position (along the mid-plane of the body). Later (*center*) the head lies sagittally and the shoulders transverse or diagonally. As it comes out the head flexes strongly backwards. Finally (*below*) the head, once it has emerged, shifts transversely, and the shoulders turn to a sagittal position to come out

149

her hands, clearing its nose and mouth so that it can breathe better and freeing it from the umbilical cord if this is wound around its neck. Birth in primates is a solitary event, with no outside help. However, in humans the mother cannot see the newborn's face because it faces in the opposite direction, and any attempt to pull it could, given the extreme dorsal flexion of the head, damage the spinal cord. Because of this women in all cultures seek assistance at the moment of birth; human birth is a social activity rather than a solitary behavior.

We have two australopithecine pelvises in a reasonably good state of preservation: that of Lucy, and that of Sts 14 (an *Australopithecus africanus* from Sterkfontein). Both belong to small individuals, and are therefore probably female. In Lucy's pelvis the entrance to the birth canal is very wide in the transverse direction. This could be a trait of the species or simply an individual character, since in Sts 14 the entrance to the birth canal is more rounded (in some women the birth canal is very wide from side to side, a condition known as *platypellic*). In any case, in Lucy the fetus' head would have lain transversely, accommodating itself to the form of the pelvic inlet. Robert Tague and Owen Lovejoy believe that the position of the head would be the same on leaving the birth canal (there would thus be no rotation), and that the fetus' trajectory would be linear and toward the back, rather than curved and toward the front (Figure 10.1). So this would be a transverse type of birth, a form without known equivalent among primate species.

However, the morphology of the ischium and the pubis leads us, like Christine Berge, to conclude that the vagina was angled forward, not backward, in female australopithecines; this would mean that the birth process in australopithecines would be similar to that of modern humans, with rotation and a curved trajectory.

Before moving on in our discussion of birth in early hominids, let us spend a few moments considering the implications if this theory were proved true. In primates the vagina opens dorsally (toward the back) and penetration during copulation is from behind. Gibbons and orangutans, which spend much time in the trees, often mate hanging from their arms and face to face, but chimpanzees and gorillas copulate from behind. The bonobo has very varied positions, including face to face. In humans this is the position favored by the ventral (forward) position of the vulva, and it is characteristically human. We are also distinct in this, and if our conclusion about the direction of the vagina in australopithecines is correct, face-to-face copulation would also be the rule among these primitive hominids.

There has been much discussion of whether birth for Lucy would be as difficult as it is for modern women, or as simple as it is for the great apes.

In fact the latter are the exception to the rule among primates: in gibbons and other catarrhine and platyrrhine monkeys the diameters of the head of the fetus are fairly close to those of the birth canal. This gives rise to a certain number of pelvic–cephalic conflicts which result in the death of the fetus, the mother, or both. This is an unfavorable consequence of the fact that anthropoids have large brains in proportion to their generally small body size.

In women the pubis is longer (in relation to the overall size of the pelvis) than in men. As the pubic bone forms the anterior wall of the birth canal, this lengthening of the pubic bone in women produces a wider birth canal. Lucy and Sts 14 have an extraordinarily long pubis, and thus a much larger birth canal, in relation to body size, than modern women. The extremely elongated pubis in Lucy and Sts 14 gives rise to a pelvic morphology known as *hyperfeminine* according to modern standards. These and other characteristics lead us to conclude that the pelvis of australopithecines was already designed to resolve the problem of birth that bipedalism had created. Although this problem was perhaps not as acute as in our species, the advantages enjoyed by female chimpanzees, gorillas, and orangutans had been lost.

We can consider this question of the rhythms of development right from the moment of birth, since among apes the newborn comes into the world with a brain more than one-third of the volume of the adult brain, while in our species it is less than one-quarter. This means that at birth the newborn human is much less developed cerebrally, and therefore more helpless, than any other primate. Primates in general are a group of mammals which give birth to few offspring at a time (usually only one), but these are well developed when they are born. This developmental precocity is also found among apes, so the less developed state of our newborns must be a condition our evolutionary line has acquired since the time of any ancestors whose babies were more developed.

We have no direct information as to the size of the newborn australopithecine's brain, since no neonate skull has been found, but we can approach the problem in another way. The pelvises of Lucy and Sts 14 tell us the dimensions of the birth canal through which the head and shoulders of the fetus would have to pass during birth. The dimensions of the birth canal establish the maximum size of the skull and the brain of the newborn; in the case of Sts 14 and Lucy this would be comparable to that of a newborn chimpanzee, or perhaps a little bigger. Furthermore, the brain size of the average adult *Australopithecus afarensis* (Lucy's species) and *Australopithecus africanus* (the species of Sts 14) would also be approximately the same size as that of a chimpanzee (about 400–410 g), or perhaps bigger. Thus, the ratio between the brain size of the newborn and that of the adult (which indicates the degree of cerebral development of the newborn) would be comparable in the great apes and in the first hominids. So the newborn babies of the first hominids would have come into the world much more developed than modern human babies, although the birth process was already like that of modern humans because the australopithecines were bipedal like us.

We have already seen that in *Homo ergaster* body size had increased substantially in relation to australopithecines, *Paranthropus*, and *Homo habilis*. But at the same time the brain volume of the adult doubled. The brain of the newborn may also have doubled in size, leaving the proportion between the two brain sizes the same. But it is also possible that the dimensions of the birth canal did not increase by the same amount; in this case the brain of the fetus could not be as large (since it had to pass through the birth canal), and it would represent a smaller percentage of the adult brain than in australopithecines. This would imply that the newborn came into the world less developed, needing more care.

Although this supposition accords with the generally held view (which we shall discuss later) that *Homo ergaster* already exhibited major changes in the social sphere compared to australopithecines, and we imagine a more protective social environment, we have no definitive proof that young *Homo ergaster* were born less mature. Unfortunately, the only pelvis of the species we have is that of the Turkana Boy, which in addition to being that of a child is very incomplete and allows us to draw no firm conclusions on this subject. Nevertheless, it is worth mentioning that Christopher Ruff and Alan Walker have estimated, on the basis of the Turkana Boy, that the newborn of the species would have a brain one-quarter the size of the adult brain. Bearing in mind that any such conclusion is as yet unproven, this would nevertheless imply that the newborn *Homo ergaster* was already less mature.

Childhood and Adolescence

Having established that the newborn australopithecine would have been no more developed than newborn great apes of today, we need to determine how long the different life stages lasted in the first hominids.

When he published *Man-apes or Ape-men* in 1967, Wilfrid Le Gros Clark thought that the development of australopithecines and

Paranthropus was similar to that of modern humans, slower and of longer duration than that of the apes.

Fortunately, we have a way of establishing the age at death in some cases, based on analysis of the lines of growth of the tooth enamel. Enamel is deposited in layers as the crowns of the teeth form; if we cut a cross-section through the enamel, the boundaries between the layers can be distinguished as what are known as striae of *Retzius*, clearly visible under a microscope. Each of these lines corresponds to the end of a cycle of deposition of enamel lasting approximately one week. By counting the number of striae of Retzius in a tooth where the development of the crown is not complete, or was completed shortly before the individual's death, we can work out how many weeks elapsed from when the tooth began to form. If we know when the formation process began, we can determine how old the individual was when he died. We do not need to cut through a fossil tooth to observe the striae of Retzius, because they can be seen on the surface.

This method of calculation has been applied to the incisors of some hominids in which the crown is newly formed or has recently finished growing. There is a fossil of *Australopithecus africanus* (Sts 24, from the Sterkfontein deposit), and another of *Australopithecus afarensis* (L.H.2, from Laetoli), which show a stage of dental development similar to that of the Taung Child. Remember that at the beginning of this chapter we established the premise that if the Taung Child died between the ages of three and four, we would conclude that development in his species (*Australopithecus africanus*) was like that of chimpanzees. In fact, when the method of lines of enamel deposition was applied to incisors of Sts 24 and L.H.2, the ages of death obtained were between 3.2 and 4 years. Tim Bromage and Chris Dean, who carried out this investigation, also obtain ages close to those of chimpanzees for the emergence of the first molars in *Paranthropus*. Everything thus seems to indicate that Le Gros Clark was mistaken and that the life stages of australopithecines and *Paranthropus* were no different in duration from those of the great apes.

The method of counting the lines of enamel cannot be used on the Turkana Boy's incisors, because he was already too old when he died, and we thus cannot say with certainty at what age he died. If we assume development patterns similar to those of modern humans, he would have died at the age of 11, while if his development was like that of chimpanzees and gorillas, he would have been about 7 years old.

Fortunately, there is an indirect (although less precise) way of approaching the problem. It has been observed that in all primates brain size is very closely correlated with longevity and with the duration of the different life stages. The greater the average brain of a species, the longer the life of the individuals of that species and, for example, the later the second adult molar appears. The brain volume of *Homo ergaster* is thought to have averaged between 800 and 900 cc, midway between that of chimpanzees (and australopithecines and *Paranthropus*) and our own. On this basis we can, in principle, assume that the rhythm of the development of the species would also be roughly midway between the two, and that Turkana Boy died at the age of nine or ten. If this is the case, it would mean that a significant change in the biology of development had occurred in this species compared with australopithecines and *Paranthropus.*

There is one further aspect we need to consider in order to complete this look at development. Turkana Boy is estimated to have been about 160 cm tall. However tall he might be when fully grown, this height would be too tall for a modern child at his stage of development (as we have noted, equivalent to about 11 years old). In other words, the height of Turkana Boy would be more suited to a modern adolescent of 15 or older (even among very tall populations such as the Maasai).

However, in our species an acceleration in growth occurs at an age older than that of Turkana Boy – the so-called "growth spurt" of puberty (around the age of twelve in girls and fourteen in boys). The conclusion drawn by Holly Smith, who has studied the development of Turkana Boy, is that in *Homo ergaster* this change in the rate of growth did not yet exist, and that it only occurs in our species. In apes and other primates development is more continuous, without any marked accelerations in adolescence, and the percentage of growth remaining is lower in apes of an equivalent "physiological age." No doubt this was also true of *Homo ergaster* (though even so, Turkana Boy would have grown to be over 180 cm tall).

11

Social Intelligence

As long, therefore, as we followed these reflections, we could not but conclude that man is by nature fitted to govern all creatures, except his fellow-men.

Xenophon, *Ciropaedia*

The Unexciting Sex Life of the Female Orangutan

Five years after her last period of sexual activity, the female orangutan comes into season once more. During this long break she has been gestating, giving birth to and nursing her latest offspring. Now the moment has come for weaning and beginning a new cycle, a new gestation. An ovum awaits fertilization, probably by the same male as the previous time. After this brief receptive period, the sex life of the female orangutan ceases for another five years or so (unless the baby she has conceived miscarries or dies).

Female chimpanzees and gorillas also have seasons (known in zoology as periods of *estrus*) separated by breaks of several years, generally more than three years and less than six, but while female gorillas have sexual relations with only one male, female chimpanzees will have several lovers. Human women, on the other hand, do not have a period of estrus, and thus the moment of ovulation cannot be detected; unlike female apes, their sexuality is not governed by being in season and it includes the long infertile periods of gestation, lactation, and menopause. In other words, whereas female sexuality in apes is linked exclusively to reproduction, in our species it also exists beyond this function.

What causes the differences in sexual behavior, and by extension in social behavior, in species that are so closely related? The answer is genes – the same factor that causes male orangutans and gorillas to be much more heavily built than females, and the males of these species to compete amongst themselves to form harems. In chimpanzees and humans there is less difference in size between the sexes, and in neither case do the males fight one another to establish groups of females (see Figure 2.3). Just as some genes determine our physical characteristics, others program our behavior, and both kinds are subject to the harsh test of natural selection over the course of the generations.

Behavior as Adaptation

With their research, Konrad Lorenz and Niko Tinbergen (who shared the Nobel Prize for Medicine with Karl von Frisch in 1973) laid the foundations for understanding animal behavior, creating a new scientific discipline – *ethology*, or the study of behavior. Ethology establishes that there is a degree of genetic programming which codetermines behavior (in other words, it determines it only in part). Just like morphological structure and physiological function, behavior has to adapt to the way of life of the individuals of different species, and therefore the genes which determine it are subject to natural selection.

Ethologists have demonstrated that many of the behavior patterns of animal species are innate, and moreover that they develop, like the organs, over the life course of individuals. Thus young animals have certain infantile behaviors which are only useful for surviving at this stage of life, when they are strongly dependent on their parents, in a situation of direct competition for food with other young. Complex behaviors of courtship, mating, and care of offspring, however, mature at the same time as the reproductive organs, and in many cases develop even if the animal has grown up in extreme isolation, reflecting their innate rather than learned nature.

Konrad Lorenz (1903–89) discovered that the geese he studied at his family home in Altenberg were programmed to recognize as mother the first moving object the goslings saw at the moment when they hatched. In normal conditions this would be the true biological mother, but they can be made to adopt as mother a person or even an inanimate object which appears in front of them at the decisive moment. Thus the different signals which trigger behaviors in animals can be broken down and analyzed, making ethology an experimental science.

Perhaps the reader is surprised or disappointed to find that there are genes underlying our behavior, just as there are for the color of our eyes, or belonging to one or other sex. But what science requires is not acts of faith, but experimental proof of hypotheses, and today there is no doubt that there is an element of genetic determinism in behavior. In any case, would it be preferable if human beings came into the world as a "blank slate," with nothing written on it? If learning processes were the only thing which determined our behavior, would it not be much more terrible to be in the hands of those who have the power to program education? How could we be free if we are entirely conditioned by the education we have received?

Of course, ethology does not oblige us to take the reductionist course of believing that all our behavior is planned, from the cradle to the grave, and that we have no capacity to make our own decisions. In fact, programming of this kind would not be very adaptable because each individual lives in his or her own ecological and social environment, and has to adapt to it. An ant is much more strictly programmed in its behavior patterns than a mammal. Humans are an extremely intelligent species of social primate, and we have great flexibility in our behavior, allowing us to respond differently, on the basis of our own experience or education, to the different situations we face in our environment. Life throws up many unpredictable problems, and the solution cannot therefore be in the genes.

We shall return to this subject at the end of this chapter, because one decisive factor in the expansion of our brain seems to have been the need to analyze and take decisions in relation to a particularly changeable and unpredictable aspect of our environment – the behavior of other members of our group. As social primates living in large communities, we need to process a large quantity of information on an enormously complex system (the community) that comprises many elements (individuals) relating to one another in a virtually infinite number of ways.

With the development of information technology in today's world, it is easy to understand that the more instructions a computer has, the more software that is loaded into it, the greater its flexibility and capacity to do different things – including being more efficient at analyzing situations and taking decisions. Very soon computers will even be able to learn from their own experience. In other words, genetic programming is not an enemy of freedom, but allows us to evaluate the different options and choose between them.

At the beginning of this book we noted that the way an individual uses or does not use its organs during its own life has no effect on the organs of its descendants (despite what Lamarck maintained). Similarly, the information accumulated by an individual about his environment during his life cannot be transmitted through the genes. However, this wealth of useful knowledge does not necessarily have to be lost, since it can be transmitted between generations by non-genetic methods, through education. In the case of humans, the passing of information from one generation to another is called *culture*; this type of permanent collective memory is partly universal and partly varies with each ethnic group, each community, each family. Because it is cumulative, it has made possible the great advances in science and technology. Lorenz's goslings had no genetic programming to tell them exactly what their mother looked like, but they had some simple rules telling them how to find out for themselves. Similarly, we humans have an innate disposition for learning a language from a young age, but our genes do not program us to learn English, Spanish, or Chinese.

Having made this foray into the principles of ethology, let us return to the social life of our closest relatives and attempt to enter into this aspect of biology which does not fossilize. In this area we owe a great deal to the pioneering work done by Jane Goodall with chimpanzees, Dian Fossey (1932–85) with gorillas, and Biruté Galdikas with orangutans.

Comparative Sociobiology of Hominoids

All anthropoids (Old and New World monkeys, apes, and humans) are social, with one exception. The term "social species" is used to refer to those in which lasting relationships between adults are established. The only exception to this rule are the orangutans, which are solitary animals in that the only stable bonds are those between mothers and their non-adult offspring. Adult males and females only come together during the brief and very widely spaced periods of estrus of the females. There are no relationships or alliances between males or among females. Each female lives in her small territory with her offspring, and the wider territories of the males encompass those of several females, with whom, however, they only come into contact to reproduce. Males compete with one another for territory and for the females within it, and this competition results in them becoming much heavier, twice the weight of females. It could be said that male orangutans form harems, but these are harems in which the females live dispersed rather than all together.

Gorillas, on the other hand, are highly social. As their food supply is abundant and constantly available they do not need large territories and they do not travel great distances in one day. Each group of gorillas is formed by an adult male, the silverback, and his harem, a group of females with their offspring, all descendants of the silverback. When a female or a male reaches puberty they leave the group. Males compete with one another for females, and this is why they are so big, although the difference between the sexes (known as sexual *dimorphism*) is less marked than in orangutans: the average female gorilla weighs about 60 percent of the average male, or a little less than two-thirds.

Common chimpanzees do not form harems. When females reach adulthood they generally leave the group; however, when males reach adulthood they stay in the group. This means that all the males in a group of common chimpanzees are related, while the adult females are not. Each community controls a territory, which the males defend against other alliances of males. These struggles between groups of related males are violent fights, sometimes to the death.

The trees which supply the fruit eaten by chimpanzees are dispersed, and the fruit ripens at different times; for this reason within each territory there are moments of *fusion*, when many individuals gather around a tree with ripe fruit, and of *fission*, or dispersal in search of sources of less abundant resources. When a female is in season, marked by a spectacular anogenital tumescence (swelling of the zone around the sex organs), there is no competition among males; rather, several of them mate with the female at different times. Because of this promiscuity and the absence of harems the males are not much bigger than the females, who on average weigh at least 80 percent, or almost four-fifths, of the average male.

In discussing the behavior of apes we cannot emphasize too strongly that we are talking about primates, not insects; thus their rules of behavior are very varied and any attempt to generalize them is a blatant simplification. For example, sometimes male chimpanzees at the top of the hierarchy have been observed to push in front of those lower down the hierarchy when mating with females in estrus; there are also cases where male–female couples form, lasting several days, or even the full two weeks of estrus.

Genetic analyses of the relationship between individuals in chimpanzee communities are currently underway; in the future these will throw a great deal of light on how all of these behavioral strategies are reflected in reproductive success. In order to make these "paternity tests," hairs which have fallen out in the "nests" the chimpanzees build in the trees to sleep, and mouth cells left on bitten fruit, are used. On the basis of this method, Paul Gagneux and his colleagues have observed that females frequently "escape" from their community and are fertilized by a male from another group.

In pygmy chimpanzees (bonobos), the bonds between related males are not as strong, but there are closer relationships between females, even though they are not genetically related. Curiously, these interfemale relationships prevent the isolated males from imposing their hierarchical authority. On the other hand, the males do not appear to take defense of the community's territory very seriously. Sexual dimorphism is similar to that of the common chimpanzee.

Natural Selection and Sexual Selection

When Darwin spoke of natural selection he was referring to the elimination of defective individuals and the survival of the best adapted. Spontaneous variations produce new types of organism, often unviable, but sometimes with characteristics which allow them either to cope better in their way of life (to occupy their ecological niche better, as we would say now), or to exploit resources that their competitors do not use (expanding or changing their ecological niche). This is how new species appear.

However, Darwin also realized that in many species the males showed characters which were not adaptive from the ecological point of view. These characters make the males more showy, or stronger (as we have seen in the case of gorillas and orangutans), sometimes giving them weapons to fight other males of their own species (and only secondarily useful in struggles with animals of other species). In order to explain this apparent exception to his theory of natural selection, Darwin developed the theory of *sexual selection*. Briefly, this means that females select the most decorated male, or passively accept the one who defeats all the other males, showing in either case that he is an individual with excellent health and strength, and thus the best possible progenitor among those competing. Darwin put forward this theory in his book *The Descent of Man, and Selection in Relation to Sex* (1871). As he put it, sexual selection "depends on the advantage which certain individuals have over others of the same sex and species solely in respect of reproduction." However, Darwin could not have imagined that competition was established at a level below that of the individual – at the level of the sperm.

The great primatologist Adolph Schutz (1891–1976) observed in 1938 that primates vary greatly in the size of their testicles relative to body weight. For example, a male chimpanzee weighing 45 kg has testicles weighing approximately 120 g (60 g each), while a male gorilla of 160 kg has testicles which together weigh 30 g. Schultz did not know at the time how to interpret this difference, but the modern anthropologist Alexander Harcourt and his collaborators have put forward a very original hypothesis.

Comparing the relative weight of the testicles in different genera of primates, it can be observed that the species in which the males have very large testicles are those in which the social groups include several males, such as chimpanzees, baboons, and macaques. Gorillas, on the other hand, which live in groups where there is only one male, have relatively small testicles. The size of the testicles appears to be related to the quantity of sperm they produce, and also to the length of the tail and the mobility of the sperm. Harcourt's hypothesis is that in species with large testicles, a female may be inseminated by several males when she is in season, and the sperm of the males competes, in quantity and quality, to fertilize the egg. Male gorillas, however, do not have to compete in terms of sperm, because they do not let any other adult male approach their group of females. Gibbons, which are strictly monogamous, orangutans, which are polygamous but not social, and humans have a testicle weight no higher than normal for a primate of their size.

In the social categorization of primates, humans are usually described as monogamous – a definition which might cause some readers to shake their heads! Admittedly there is wide variation in the human family structure in different cultures (which have little to do with the social context in which our evolution took place). It is obvious that gibbons are monogamous and do not form groups, that orangutans are polygamous and live separately, that gorillas are polygamous and form harems, and that chimpanzees are social and promiscuous, but surprisingly, we are not as sure about the social biology of our own species.

From the biological point of view no specific answer can be given. However, one thing is certain: humans form social groups which include many male individuals, so it would seem appropriate to place us in the group of primates in which there is selection among sperm, like the chimpanzees. In strictly zoological terms, the fact that this is not the case means that we are a species in which it is rare for a female to have sexual relations with several males during the period of ovulation. Moreover, as ovulation is not signaled overtly as it is in other primate species, males have no way of knowing when it occurs, so we can simplify the previous sentence to say that it is rare for a female to have sexual relations with several men. Furthermore, the level of sexual dimorphism in the body weight of our species (around 83 percent) indicates that there is not a high degree of competition for females among the males.

In other words, we are a special type of primate, and we see male individuals living together in society, but also a degree of exclusivity in the sexual relations of each man with a woman, at least for a period of time. In other words, something like monogamy.

But having set out what we know and what we do not know about the social biology of our species and other species of primate, it is time to investigate how the first hominids behaved in society.

Bipedal and Monogamous from the Beginning?

Rob Foley observes that there are no examples among apes of groups based on related females (whereas this is common among Cercopithecidae, or Old World monkeys: only in the red colobus monkey are there relationships between related males). He concludes that the social structure found among chimpanzees, involving alliances between related males to defend a common territory, and dispersal of adult females outside their birth territory, must also have been the rule among the first hominids, whose social biology would initially have differed little from that of chimpanzees.

However, australopithecines and especially *Paranthropus* lived in more arid environments than modern chimpanzees, in sparser woods where the sources of food would have been more dispersed and less abundant. This leads Foley to believe that australopthecines and *Paranthropus* would have had larger territories than chimpanzees. They would probably have maintained alliances between related males, both for defense of resources against other coalitions of males, and against predators, which would have become more dangerous as the tree cover decreased. However, smaller social units would have formed within these large territories, since the resources available in the environment would not have allowed all members of a given community to be together at all times. Thus, the social system would also have been one involving fusion, with much of the group gathered around a very abundant source of food, to travel long distances over open ground, or to sleep, and fission, with the group separating into smaller units to find food during the day.

Determining what these smaller units might have consisted of is another question. Jane Goodall and her colleagues have studied the spatial distribution of a population of just under 150 chimpanzees in Gombe National Park, in Tanzania, since 1960. The most closely studied community in the park was made up, over the years, of between 4 and 13 adult males, 8 to 18 adult females and 18 to 31 immature individuals. The group's territory ranged from 6.75 to 15 sq km, occupying between

3 and 6 valleys in the central region of the park. The adult females spend more than 65 percent of their time alone with their young, feeding in their own small areas of around 2 sq km, which partially overlapped with those of others. The adult males are more social, and travel throughout the community's territory, patrolling and defending their borders together. This type of spatial distribution, with females spending time alone with their young, is unlikely for australopithecines, even more so for *Paranthropus*, because of the risk from predators in a more open environment.

However, the family units might have been comparable to those of hamadryas baboons (*Papio hamadryas*) and geladas, i.e., formed by one male with a number of females (and their young); these may occasionally have joined together with other small harems to form large groups to find food, travel, or sleep. This is a social system which we know to function successfully in ecosystems similar to those of our ancestors, like the arid environments where the hamadryas baboons live. But one author, Owen Lovejoy, believes that australopithecines were monogamous, and moreover that monogamy is closely related to the development of the hominids' bipedal posture. Let us look at his argument.

Monogamy, or pairing off for reproduction, is not found only in humans. It is also observed in many birds and primates, including gibbons. But as we have already noted, our species has the additional unusual feature of a permanent sexual relationship, for most of the time without reproductive function. To put it even more clearly, defining human sexuality purely in terms of procreation is not natural (in the biological sense), but the absolute opposite. Among humans sex also exists to keep the couple together; in other words, it is at the service of love.

However, this romantic function of sex does not contradict Darwinist principles, but in fact reinforces them. The long period of development of human young makes it impossible for a mother to look after several offspring at once (in the context of a hunter-gatherer economy). The stable couple, in monogamy, means that the father is involved in the task of supporting the family, which operates as an economic as well as a reproductive unit.

There has been much debate as to what is understood by the contribution of fathers (males) to the raising of young, how this contribution can be measured, and to what extent it occurs in the different primates and human societies. Whatever the case, the situation in our species bears no relation to the habits of chimpanzees, gorillas, and orangutans (our closest relatives in other senses), where males have no connection with their offspring; all we can say is that among chimpanzees and gorillas (not orangutans, which are not social), the males are tolerant of the young and protect them against predators and being killed by other males (a serious risk, as we shall see).

Lovejoy maintains that bipedalism has nothing to do with our ancestors' adaptation to open environments, as has so often been suggested, and that it probably arose when we still lived in the forest. Bipedalism, according to Lovejoy, was not related to temperature regulation, to efficiency of locomotion, or to freeing the hands to make tools. On the contrary, bipedalism would have freed our hands and arms to carry food. In Lovejoy's view males transported food in this way to the base camp to feed the females and their young, who thus avoided many dangers by not having to travel with their mothers.

Lovejoy even believes that this made it possible to reduce the long period between births characteristic of chimpanzees, gorillas, and orangutans, and to increase the number of young over the fertile life of the females. This supposed advantage of the first hominids is highly debatable, because even in modern human hunter-gatherer societies (i.e., pre-agriculture and pre-animal husbandry) the interval between births is still long, from three to four years.

We are used to seeing natural selection from the point of view of individuals, but in order to understand what follows we need to look at it from the point of view of genes. As Richard Dawkins noted in his famous book *The Selfish Gene*, individuals die, but genes remain. Or more correctly, they are preserved as copies in other bodies, those of our children. Dawkins takes his argument to the extreme, affirming that genes use us for their own benefit and even sacrifice our bodies if necessary, as in the case of what is known as altruistic behavior, when parents put their own lives at risk to save their children. In fact, the time and energy expended in raising offspring, although less heroic, can also be considered altruistic, since it is not aimed at satisfying the interests of the parents, but those of the children. Everything has a point according to the logic of natural selection, because the genes of parents who do not behave altruistically and abandon their children (thus dispatching them to their death) will not be present in the succeeding generation.

These explanations enable us to understand behaviors which appear monstrous if we view them in moral terms. We noted earlier that when they reach puberty, female gorillas and chimpanzees leave their native territory and their community to enter another which is completely alien to them, and into which they are nevertheless accepted. However, we are talking here of females without young, because if they appear with a nursing infant in the alien group it is virtually certain that the baby will be killed. The dominant male in the case of gorillas, or the related males in the case of chimpanzees, have no interest in the genes carried by the newly arrived young individual, but they are very interested in its mother being available as soon as possible to be fertilized. Interrupting lactation by the rapid method of infanticide places the mother in a situation where she can begin a new ovarian cycle. The reader can imagine what happens to the young in a harem of gorillas when the silverback is replaced by another male (as a result of natural death or defeat in a fight). The logic of the genes is implacable.

If we follow this logic, a male of the first bipedal hominids providing food for a female with young, as Lovejoy suggests, would have to be sure that the young carried his genes. If the females of the species had periods of estrus they would have to be watched closely during the whole of this period. On the other hand, if the female had no estrus, making it impossible to know when she was ovulating (and thus monopolize her during this period), the only viable alternative to ensure paternity was monogamy and sexual fidelity.

In fact menstruation may have evolved as an indicator of fertility, because a few days after it has occurred there is a fertilizable egg. However, Beverly Strassmann points out that other primates do not appear to make these calculations, since chimpanzees (like macaques, Old World monkeys, and baboons), which menstruate copiously, indicate ovulation to the males by means of swelling of the sexual organs.

As we pointed out above, social life is not preserved in the fossil record. But we also noted that among gorillas and orangutans, males compete with one another for females, and that as a result there are great differences in body size between the two sexes, whereas the two sexes are more similar in chimpanzees and humans (although there are still differences; in gibbons, which are strictly monogamous, there is no sexual dimorphism of weight). If this is a universal rule linking anatomy to social biology, it could offer a key to approaching the problem in extinct species. This type of exercise, which consists of applying relationships observed in the modern biosphere to fossil species (with the premise that such relationships have always existed), is known as *actualism*, and is one of the tools most frequently used to study life in the past.

Australopithecus afarensis, the oldest hominid species of which there is sufficient fossil record to study the difference between the sexes, shows great variation. So much that some authors saw two or more distinct species among these fossils; it was Tim White who grouped all the fossils into a single species, as we have already noted. In fact this is a species which shows great sexual dimorphism of size between males and females (around 66 percent of body weight, very close to that of gorillas), as Henry McHenry, Charles Lockwood, Brian Richmond, William Jungers, and William Kimbel, among others, have highlighted (Figure 11.1). From this point of view it does not seem that Lovejoy's hypothesis, that bipedalism was linked to monogamy from the beginning, is tenable (if this were the case we would expect much less marked sexual dimorphism in *Australopithecus afarensis*). A social system closer to that of hamadryas baboons and geladas, as suggested by Rob Foley, appears more likely for the first hominids (Figure 11.2).

But there is one small problem with this interpretation. As Michael Plavcan and Carel van Schaik have shown, in modern primates the length of the crown of the canine is an even better indicator than body weight of the level of competition among males for females. Now, as a result of the reduction of the canines in *Australopithecus afarensis*, the difference in the length of the canines between males and females is less than that in gorillas and orangutans, and also less than in both species of chimpanzee: it is still smaller in ourselves.

If male *Australopithecus afarensis* competed for females, why would they have had to give up their best weapons? The reduction of the canines may be related to the change in diet experienced by these hominids, since large canines impede lateral movements of the jaw. Moreover, in view of the intense wear to which the canines of australopithecines and *Paranthropus* are immediately subject, we might wonder what would be the point of having large canines which would barely project at all, precisely at a time when they would be more useful as weapons. To sum up, the absence of major sexual dimorphism in the canines among the first hominids appears to have been imposed by the type of chewing, rather than indicating a lack of competition among males for females.

Nevertheless, according to Tim White the beginnings of reduction of the canines are already observed in *Ardipithecus ramidus*, which is not thought to have had a very different diet from chimpanzees, although it might have been sufficiently different to explain this reduction. In fact, this combination of major sexual dimorphism in body size combined with very little sexual dimorphism in the size of the canines, characteristic of hominids at least since *Australopithecus afarensis*, is not found in any modern primate species, and comparison is therefore impossible.

Thus we have no conclusive answer to the question of when modern patterns of social biology appeared in human evolution. Current thinking, which tends to consider australopithecines and *Paranthropus* as "bipedal chimpanzees," vegetarians with no stone tools, no major increase in brain size, no language, and a short period of maturation, suggests that this great change took place within the genus *Homo*.

Figure 11.1   Male and female *Australopithecus afarensis*

Figure 11.2   One male and two female *Australopithecus afarensis*

Brain Size and Size of Social Group

In a previous chapter we investigated the use of being bipedal, or what kind of adaptation this is and what ecological niche it relates to. We have already seen that there is no easy answer to this question. However, no one wonders what the use of being intelligent is. We are so convinced that intelligence is a gift which makes us superior to any other living form that we do not concern ourselves with its adaptive value. Nevertheless, enlargement of the brain is a specialization like that of any other organ, and natural selection favored it because it offered advantages in the context of the ecological niche occupied by hominids, in those in which it occurred (not all of them, as we have seen). What were these advantages?

There are two moments in human evolution where a marked increase in brain size occurred which could be related to significant changes in social patterns. The first of these increases occurred with *Homo ergaster*, in whom brain volume increases from approximately one-third of the average in modern humans, as in australopithecines and *Paranthropus*, to two-thirds (*Homo habilis* had a brain size somewhere in between). The second great enlargement occurs during the last half-million years, and produces the enormous brains of our species and the Neanderthals.

As we have seen, the increase in brain volume implies a change in diet, because it affects a tissue with high energy consumption. As a result, animal proteins and fats become incorporated into the diet in substantial quantities. Unlike some of the very abundant (but much less energy-rich) plants, these resources are not continuously distributed in the environment, and are not easy to obtain; thus the size of the territory to be covered and the time spent searching for them increase. At the same time, from *Homo ergaster* onward rates of growth of young are already close to our own, implying a longer period of dependence than in apes and earlier hominids. All of this means that it would be difficult for a mother alone to take care of a number of young at the same time. It is therefore possible that the great social change took place in *Homo ergaster*, although some authors maintain that it took place during the second great cerebral expansion, that of humans and Neanderthals.

But there is more than just an indirect relation between increase in brain size and relations between the sexes: it is possible that brain enlargement is directly related to an increase in social complexity. In the first place, it has been observed that mammals which live in complex societies (such as anthropoids and dolphins) have larger brains than solitary mammals of similar size. Aiello and Dunbar have also discovered that among the

169

different species of primates, the relative size of the neocortex in relation to the rest of the brain is directly related to the size of the social groups formed by these species.

However, no such relationship has been identified between relative neocortex size and way of life, making the "ecological theory" of the origin of our highly developed intelligence less attractive than the "social theory." Nevertheless, we must keep in mind the fact that from the first *Homo*, hominids entered an ecological niche which was entirely new for primates, that of carrion eaters and hunters; this may also have favored intellectual development.

To sum up, the theory that expansion of the brain and increased intelligence (or at least a substantial part of this expansion) represents an adaptation to social life, an environment in which individuals have to both cooperate and compete with other individuals, is very plausible. An intelligence developed on this basis ("social intelligence") could easily be applied to other types of complex situation. In order to do well in this difficult social environment various tactics need to be used, from the formation of alliances with other individuals, based on genetic relationship or interest, to deception. Anne Pusey, Jennifer Williams, and Jane Goodall have observed that even among chimpanzees it is important to be born into a "good family." Although female chimpanzees often gather food alone, and do not seem to have a marked hierarchy among themselves, the offspring of high-ranking females have better life prospects than those of other females, probably because they have access to better sources of food.

The skills required for what Andrew Whiten and Richard Byrne have called "Machiavellian intelligence" (in reference to the fifteenth-century writer Niccolò Machiavelli, whose book *The Prince* gives advice on how to triumph in politics through the use of hypocrisy and unscrupulous means) include, of course, a good memory, in order to remember exceptionally complex social organograms (who is who). In addition, at least among humans, there is a certain capacity for intuiting the intentions of another and anticipating his actions, as well as the capacity for creating mental representations of hypothetical situations (not only remembering past situations), evaluating them, and acting on the result – in other words, a capacity for thinking.

Summary

Here ends the part of human evolution that took place exclusively in the African continent – our birthplace, and the place where we literally took our first steps. It is time to pause for a moment to look back, and sum up what we have discussed and draw the main conclusions from it.

Darwinism, in the updated form of the modern synthesis, is the tool which evolutionists of the end of the second millennium use in their analyses. Although this is not a closed, dogmatic system and it is open to criticism, improvement, and amendment, it forms the basis for our current understanding of the phenomenon of evolution. One important aspect of Darwinism is the adaptive nature of its basic mechanism, natural selection, which does not appear to impose any specific direction on evolution. But in this book we are describing, on the basis of the fossil record, how human evolution arose; not until the final chapter will we discuss whether this has any meaning, whether it corresponds to some Law of Evolution.

After setting out the terms of the problem (what happened, how it happened, and why it happened), we have opted to follow the advice of Chilo (usually attributed to Socrates): "Know thyself." We are primates, more specifically anthropoids, and as such we are basically visual, intelligent mammals, diurnal, tropical, living in forests and trees. Many of our morphological, physiological, and ethological characteristics match this ecological definition of the group. The fact that humans, and to a lesser extent other primates, now live in climates, regions, and ecosystems very different from the backdrop to our evolution is still an anomaly, and a very recent phenomenon in terms of the long history of the primates. In this book we have briefly recounted this history, particularly that of the hominoids, the class of anthropoids to which we belong.

Summary

Hominoids originated in Africa at least 23 million years ago, and they were numerous and varied in that continent and later also in Asia and Europe. After reaching their maximum diversity around 10 million years ago, hominoids began to decline rapidly. Most species had already disappeared by 7 million years ago, and in the present day the only really abundant and widespread species is our own.

There are two reasons, perhaps related, for the decline of the hominoids: the loss of habitat caused by ecological changes, and competition with the other great group of anthropoids, the Old World monkeys or Cercopithecidae, which are now more varied and abundant than hominoids. The great rain forests and monsoon forests have been retreating as a result of a global change in climate, with an increasingly marked cooling of the planet over the last 4 million years. This climate change appears to be caused essentially by astronomical factors, combined with a particular arrangement of the continental landmasses. In addition to the change in climate, there may have been a reduction in carbon dioxide in the atmosphere during the last 8 million years, which would have favored the spread in the equatorial latitudes of the grassy plants which form, for example, the pastures of the African savanna.

It is nevertheless important to note that the first hominids, our most remote direct ancestors, did not appear in the expanding grasslands: the fossils of *Ardipithecus ramidus*, from 4.4 million years ago, seem to indicate an entirely forest-based life, with a type of diet similar to that of modern chimpanzees. It is still not known whether these hominids were bipedal. Judging from the primitivism of this species, the divergence of the lines which lead to humans on the one hand, and to the two species of chimpanzee on the other, must have occurred shortly before, perhaps only 5 or 6 million years ago. This date for the separation of the two lines is agreed by molecular biologists, who estimate it on the basis of the genetic difference between species and the inferred rate of change.

Four million years ago there was a different species of hominid living on the shores of Lake Turkana, known as *Australopithecus anamensis*. These hominids walked erect and their dentition indicates that a change in ecological niche had occurred, and that they incorporated tough plant products in their diet: in other words, in addition to fruit, leaves, and tender stems and shoots they also ate hard seeds, nuts, tubers, roots, and other underground parts of plants. The former type of plant products is found in the humid forest; the latter are typical of more arid environments.

The hominid fossils from the subsequent million-year period are found in fairly substantial numbers; these are assigned to the species

Australopithecus afarensis. The more extensive record of this species allows us to tackle an important question: which of our distinctive features came first? We have already seen that bipedalism arose before *Australopithecus afarensis*; however, in terms of encephalization, the intelligence of these hominids was similar to that of modern chimpanzees. Their period of growth also appears to have been similar, and young would not have been born more helpless, requiring more care.

All the fossils cited up to now come from deposits in the Great Rift Valley, a great fracture through the African continent with a number of branches. This is because the lake basins which formed along this fissure provide favorable conditions for the formation of paleontological deposits. But in addition, the great mountains and high plains associated with the Rift Valley make East Africa a zone of lower rainfall than the center and west of the continent at the same latitude. Because of this reduced rainfall in East Africa there are no rain forests, but rather sparser woodland, grasslands with trees and dispersed scrub (the savanna), and great prairies. It is possible that it was in these more open and arid woods of East Africa that a primate ecologically very similar to the modern chimpanzee evolved toward bipedal forms, and toward a diet which included tough plants. However, hominids soon spread westward from their place of origin, because fossil remains have been found well into Chad. These may also be of a different species from *Australopithecus afarensis*, which would mean that there had already been an early diversification of hominids.

Between 3 and 2 million years ago we find various forms of hominids. One species, called *Australopithecus africanus*, lived in South Africa during the first part of this period, and its fossils are found in cave deposits. The australopithecines did not live in these caves: their remains were left there by predators which hunted them.

Around 2.5 million years ago, the group is clearly split into two major types of hominid. One, *Paranthropus*, developed a massive masticatory apparatus, probably a specialization to process tough, abrasive plant food. *Australopithecus afarensis* is its more or less direct ancestor. Three species of *Paranthropus* are known: the East African *Paranthropus aethiopicus*, which is the oldest; *Paranthropus robustus*, found in several South African caves; and *Paranthropus boisei*, an East African form which died out a little over one million years ago. Australopithecines and *Paranthropus* showed marked sexual dimorphism in body size. They may have formed communities of several related males, each male in his turn gathering a small harem of females.

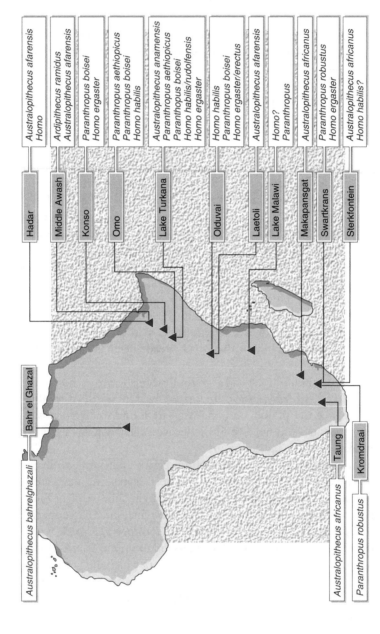

Figure S.1 Location of the main deposits with fossils of *Ardipithecus, Australopithecus, Paranthropus,* and the first representatives of *Homo*

Other hominids, the humans, developed their brain and began to manufacture stone tools. To begin with they were very similar to the australopithecines, particularly the South African form. There may have been several forms of these humans which were still similar to the australopithecines, probably spread over a large part of Africa. It is likely that *Paranthropus* and humans evolved in response to the climate change which resulted in the expansion of open ecosystems. For the first time, we meet hominoids which can be considered not strictly forest species.

Then, less than 2 million years ago, a species of humans appeared which was clearly different from all earlier hominids and from their contemporaries *Paranthropus*: this was *Homo ergaster*. Not only was their brain even bigger and organized in a different way; in addition, their face had a more modern appearance, and they were similar in height or perhaps on average taller than ourselves (the other hominids we have discussed were much smaller). Their proportions also corresponded to a body structure generally similar to our own. Their pattern of development was slower than that of the apes (and other hominids). This longer period of development implies a more protective social environment, one in which it is possible for a mother to look after several young at once; probably for the first time, males became involved in the care and feeding of young.

Two factors are crucial to understanding the enlargement and restructuring of the brain in humans. One of these, a change in diet with the regular incorporation of animal proteins, made it possible. The other factor, the increase in social complexity, gave it meaning. Intelligence developed largely as social intelligence.

These humans were capable of "inventing" a very elaborate stone-tool technology, bearing witness to their great mental capacity. Finally, they left Africa and adapted to a wide variety of land and landscapes throughout Eurasia, as we shall see in Part II.

And now that at last a highly intelligent hominid, more intelligent than any other primate but less so than ourselves, has emerged, will we see a linear evolution, a triumphal succession of increasingly intelligent species, from *Homo ergaster* to *Homo sapiens*? Or, on the contrary, will human evolution maintain its branching, complex nature, making it impossible to anticipate the end of the story at any point? Let us move on.

II

A New Home

12

New Locations for Human Evolution

Hominid forms as primitive as Pithecanthropus *and* Sinanthropus *must have lived in Europe or in the Western half of the Old World at an age still earlier than that in which advanced hominids like those of Heidelberg and Steinheim appeared.*
Franz Weidenreich, *The Skull of* Sinanthropus pekinensis

Homo erectus and the Settlement of Asia

Since the first humans were neither European nor Asian, but African, at some point in the past the first human inhabitants of Eurasia must have made their way there from Africa. When did this happen?

Since 1891, when Eugène Dubois (1858–1941) discovered a *calotte* (or skullcap) and femur at Trinil, in Java, the island has supplied a large number of human fossils. The oldest, over one million years old, are some remains from the Sangiran region and the *calvarium* of a child from Modjokerto (a calvarium is a neurocranium or braincase, a skull without the skeleton of the face). Recently geochronologists Carl Swisher and Garniss Curtiss and their colleagues, using the new laser fusion method of argon-39/argon-40 dating, have dated the Modjokerto child at 1.8 million years old, and two fragmented skulls from Sangiran (numbers 27 and 31) at 1.6 million years old. One partially preserved skull, Sangiran 4, and some jawbone fragments are probably of the same age. However, we need to treat these dates with caution, given that the relation between the fossils and the dated volcanic sediments is not at all clear (these and many other fossils from Java were collected by amateurs, rather than being professionally excavated).

Other Javanese fossils, such as the Trinil skullcap, Sangiran 2 (very similar), and Sangiran 17 (the best preserved skull of any found in Java) are between 500,000 and a little less than one million years old. Lastly, the most recent of all the human fossils from Java are the Sambungmacan calvarium and 14 more or less complete calvaria, two tibias, and pelvic remains found in the terraces of the Solo river in Ngandong.

Eugène Dubois was a Dutch doctor who enlisted as an army doctor with the aim of traveling to Java, where he hoped to find the missing link between "ape" and man, following the theories of the famous evolutionary biologist Ernst Haeckel (1834–1919), who believed that we originated in the forests of Asia, rather than those of Africa. With the skullcap and femur he found at Trinil, Dubois believed he had discovered the link he was searching for, and he named it *Pithecanthropus erectus* ("erect apeman"), known today as *Homo erectus*. All the Javanese fossils are ascribed to this species, although the Ngandong remains, the most recent, are somewhat different, reflecting a degree of evolutionary change.

The Javanese fossils such as Sangiran 4, Sangiran 2, and the Trinil skullcap itself differ little in their general architecture from the African fossils of *Homo ergaster*. Their cranial capacity is estimated at between 800 and 950 cc. In fact the two species share many primitive traits, their common inheritance from earlier hominids. For example, although the brain is much larger than in australopithecines, the neurocranium is still low (or flat), with a very broad forehead, and widest at the base, at the level of the temporal bone; from here the side walls converge toward the top, as can clearly be seen from the back.

However, the Javanese fossils show some characteristics not found in Africans of the species *Homo ergaster*, and typical of the entire species of *Homo erectus*. Essentially the neurocranium is more robust, with thicker walls, a straight and highly developed brow ridge overhanging the eye sockets, and another, also very conspicuous, transverse bony reinforcement at the back of the cranium; this is known as the *occipital ridge*. There are also some bony thickenings at the top of the cranial vault and in other areas of the neurocranium. The side view shows that the occipital bone is sharply angled.

Although there are fossils which do not belong to *Homo erectus* which show some of these traits indicating a robust neurocranium (albeit generally in attenuated form), the combination of all these features is only found in this species. Moreover, the base of the skull of *Homo erectus* shows a series of specializations.

The only fossil specimen from Java with a well-preserved facial skeleton is Sangiran 17; this has a very broad, fairly flat face, which looks like a robust version of *Homo ergaster* fossils such as ER 3733 and WT 15000. Furthermore, a jawbone fragment and some cranial remains from Java which have been very sketchily described are extremely robust, so much so that there are still some authors who seek to create a new genus, *Meganthropus*, on the basis of these fossils. Nevertheless, until we have more complete fossil documentation it is too soon to accept the coexistence of two human species in Java.

From the evolutionary point of view, it seems reasonable to accept that *Homo erectus* derives from *Homo ergaster*, although there are important changes which justify considering them as separate species. In 1961 Louis Leakey found an incomplete braincase (OH 9), dated between 1.4 and 1.2 million years old, at Olduvai. From the morphology of what little remains, this appears to represent an intermediate link between *Homo ergaster* and *Homo erectus*. However, if it is confirmed that Java was first populated 1.8 million years ago, it would mean that this specimen could not occupy such a position in human evolution.

In China two presumed stone tools, a jawbone fragment with a molar and premolar, and an isolated human incisor from the Longgupo deposit, could be of a similar age to that currently attributed to the first Javanese fossils. However, the age is debatable, and the jawbone may, according to some experts, belong to a relative of the orangutans rather than a human ancestor. The incisor is certainly human, although we do not yet have enough information on the position of this fossil in the stratigraphy of the deposit (the sequence of layers which make up the deposit).

Leaving aside the problematic finds of Longgupo, the oldest Chinese fossil is a very poorly preserved skull from Gongwangling (Lantian), which appears to be a little less than one million years old. The Chenjiawo jawbone (also from Lantian) may be of the same age or a little more recent.

The most complete human fossil record in China was found in the Zhoukoudian cave, about 50 km from Beijing. Since 1921, when excavations began, the remains of many different individuals, including two braincases, have been found. Unfortunately, almost all the fossils were lost in 1941 when they were sent, for "safety," from Beijing to the USA, under the guard of US marines. The fossils never arrived at their destination because the convoy was captured by the Japanese, although it does not appear that they succeeded in seizing the fossils; whatever the case, nothing more was ever heard of them. As a consolation we have the magnificent study made by Franz Weidenreich (1873–1948), and casts of the fossils.

181

More recent excavations have discovered some new human remains. The Zhoukoudian fossils were found at various levels of the deposit and are dated between 250,000 years (the most recent), and 550,000 years (the oldest). From the evolutionary point of view the Zhoukoudian human fossils correspond to the same species as those from Java, *Homo erectus*. Their cranial capacity, estimated from five calvaria, ranges from 915 to 1,225 cc.

The First Europeans

Recently a human mandible was discovered in Dmanisi (Georgia), in the southern Caucasus, often called the gates of Europe. This fossil is thought to be very old, around 1.5 million years, although it may be more recent; unfortunately it is difficult to tell from a single jawbone what these hominids were like.

Although, as we have seen, many doubts still remain over the age of the first Asians, it seems certain that they are well over a million years old. When did the first humans arrive in Europe?

The oldest evidence of human presence in Europe could be that of Cueva Victoria, in the Murcia region of Spain. Here the phalange of a hand has been found, which is certainly that of a primate and could be human (it appears so to us); alternatively it might be from *Theropithecus oswaldi*, a large ape we have already mentioned, which has been identified in the same cave on the basis of a molar. Unfortunately, the phalange was picked up away from its original location, and although it has been assigned to a breccia (layer of coarse-grained rock) dated at more than one million years old, we can never be completely certain in these cases.

Also in the Iberian peninsula, Josep Gibert and his colleagues have presented as human some fossil remains from the Venta Micena deposit in Orce, in the Granada region of Spain. The deposit is a little over one million years old, but in our opinion the fossils are not human.

For many years after its discovery in 1907, the Mauer mandible (found near Heidelberg in Germany) was held to be the oldest human fossil in Europe. Its age is estimated at 500,000–600,000 years. More recently, in 1993, the shaft of a human tibia was discovered in the English deposit of Boxgrove; this was considered to be of a similar age to the Mauer fossil. Some authors came to the opinion that the first human settlement of Europe occurred a little over 500,000 years ago.

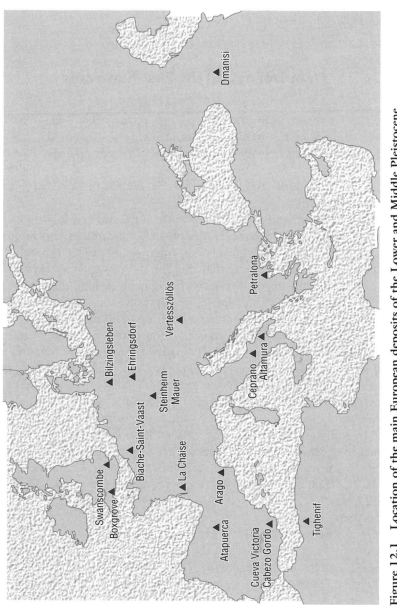

Figure 12.1 Location of the main European deposits of the Lower and Middle Pleistocene

However, the debate on the first human settlement of Europe was to change irrevocably in the summer of 1994, with the discoveries made at Gran Dolina, in Spain.

Gran Dolina and the First Europeans

The deposit known as Gran Dolina belongs to the assemblage of karstic (cave) deposits of the Atapuerca Mountains, very close to the city of Burgos, which began to be excavated in 1978 under the direction of Emiliano Aguirre. Since 1991 the project has been led by Aguirre's colleagues Juan Luis Arsuaga, José María Bermúdez de Castro, and Eudald Carbonell.

In the late 19th and early 20th centuries a railroad was built to carry mineral ores from the Demanda Mountains to the area around Burgos. The railroad route traces a great curve, entering and passing through the Atapuerca Mountains. The trench dug for the construction exposed a series of caves filled to their roofs with sediment. Among them was the cave of Gran Dolina, which contains an impressive stratigraphic sequence 18 m thick, with 11 levels numbered from top to bottom.

A deposit is like a book which is read from bottom to top, from the last chapter to the first. In Gran Dolina the excavation (over an area of 100 sq m) is progressing slowly in the upper levels, the last chapters of the book. Level 11 (or TD11), which is around 300,000 years old, has already been excavated, and excavation is about to begin in level 10 (TD10), which is very rich in fauna and stone tools, and is approximately 400,000 years old. In addition, samples have been taken from the entire stratigraphic sequence of the deposit, and a small part of the lower levels has been excavated, revealing remains of very ancient animals and some stone tools. The team investigating the Atapuerca deposits therefore knew that in the lowest levels of the deposit, the first chapters of the book, there were indications of a very ancient human presence in these mountains. Just as sometimes we cannot resist the temptation of glancing to the end of the book to find out how it finishes, the scientist who is excavating feels the same desire to know the beginning of his story. The way he does this is to make a sample survey, which is a shortcut to the deepest part of the deposit.

In Gran Dolina a survey of 6 sq m began in 1993, and reached level 6 (TD6) in the campaign of July 1994. During this and the two following campaigns, about 80 human remains and 200 tools were recovered from level TD6.

When these human fossils were found their exact age was not known, but it was clear that they were older than the Mauer mandible. This conclusion was based on the fauna found in TD6, particularly the presence of a fossil rodent called *Mimomys savini*. About 500,000 years ago this species of rodent gave way to a primitive form of the modern water rat, known as *Arvicola cantiana*. The Mauer mandible and the Boxgrove tibia were found associated with *Arvicola cantiana*, while TD6 contains *Mimomys savini*.

The beginning of the epoch known as the Pleistocene is usually situated around 1.7 million years ago (for the last 10,000 years we have been in an interglacial period of the Pleistocene, sometimes considered as a separate epoch known as the Holocene). The Pleistocene is divided into three parts, Lower, Middle, and Upper. The boundary between the Lower and Middle Pleistocene is marked by a change in the earth's magnetic field which occurred about 780,000 years ago, when the field changed from having an "inverted" polarity to the current, "normal" polarity. The Lower Pleistocene belongs to the chron of "inverted" polarity known as the *Matuyama* chron; for the last 780,000 years we have been in the *Brunhes* chron of "normal" polarity.

Geological and paleomagnetic studies carried out by Alfredo Pérez-González and Josep Maria Parés situated the change in magnetic polarity which marks the boundary between the Lower and Middle Pleistocene above Gran Dolina level 6 (TD6). Thus, the human fossils and associated stone tools are more than 780,000 years old. This evidence forces us to accept that the first settlement of Europe occurred much earlier than we had thought, although human presence in the continent may not have been as extensive or as dense as in the last half-million years.

The human fossils in TD6 represent various parts of the skeleton of at least six individuals who died at different ages. Bearing in mind the restricted size of the initial survey, it is hoped that when this level is excavated extensively in a few years' time, it will provide a very rich sample of human fossils and tools.

The association of stone tools in TD6, which includes neither handaxes nor cleavers, is classified as Mode 1. This attribution poses an interesting problem, because there is Acheulian culture (Mode 2) in East Africa 1.6 million years ago, and in 'Ubeidiya (Israel) shortly afterwards. So why are the first European fossils not associated with this later culture?

The same problem is posed by the deposits in China and Java, which similarly have no handaxes or cleavers. There are various possible solutions. Perhaps the first humans to settle in Europe and Asia abandoned the Acheulian style of carving of their ancestors; it has been suggested that the populations of the Far East may often have used another raw material for their tools, such as bamboo.

Perhaps, as Eudald Carbonell and his team suggest, the people of East Africa who developed the Acheulian industry had a more advanced culture than those who manufactured Mode 1 tools. As a result, their population increased and they forced groups who did not have the superior culture to the marginal areas of the African continent, and finally to migrate out of Africa.

Finally, the new dating we mentioned above for Java and China suggests a new argument: perhaps the first Asians left Africa before the Acheulian culture arose. The same explanation is also possible for Europe, but here we have no reliable dates of such antiquity for the initial settlement. Not far from Venta Micena, in Fuente Nueva-3, an assemblage of stone tools (without handaxes or cleavers) somewhat older than those of Gran Dolina has recently been found, and it is always possible that others, even older, may appear somewhere in Europe at some time in the future. In any case, the first European fossils of the Middle Pleistocene, such as the Mauer and Boxgrove fossils, are associated with Acheulian technology. We have to wonder how the Acheulian culture came to Europe – whether it was with a new wave of settlers, or simply that the technology traveled from Africa, passing from some populations to others without any migration of peoples or flow of genes. The problem of whether technological changes observed in a particular place imply the arrival of a new type of human is a recurrent one in prehistory, and we shall come across it again.

Prehistoric Cannibalism

In theory, the human fossils should not be in the Gran Dolina cave. In fact, human remains "should" never be found in karst deposits hundreds of thousands of years old. What "should" be found, however, and frequently are found, are bones of carnivores and herbivores, and stone tools. The reason for this is very simple. The carnivores (lions, panthers, lynx, hyenas, wolves, wild dogs, foxes) settled in caves, making their shelters and lairs there. They lived, reproduced, and died there. Bears also spent the winter (hibernated) in caves. Herbivores do not live in caves, but carnivores and carrion eaters transport their carcasses there to eat them. Human beings occasionally used caves, spending some time there, usually not very long, making and using stone tools, and then leaving them there. Humans also brought animal prey to the deposit, and ate it there. Finally they left.

In order for a human bone to fossilize in a cave something extraordinary would have to have happened. Perhaps a carrion eater, a hyena for example, might bring in part of a human cadaver or simply a bone found outside. But for the skeletons of at least six different individuals to accumulate, circumstances would have to be more exceptional, more fortunate for ourselves, although perhaps not so much for the humans in question. Because in this case, we are looking at the remains of a cannibal feast.

The human remains in Gran Dolina appear mixed with animal remains, and fairly broken up. Some show cut marks created by the edge of a stone tool which has been enthusiastically used to separate flesh from bone. It is thus clear that the bones were stripped of flesh and eaten in the cave itself by other humans. This is the oldest known evidence for this kind of practice. It is difficult to imagine that it relates to ritual behavior, and it appears in principle that human bodies were treated with no more respect than those of the herbivores with which their remains are mixed. However, the studies currently being carried out by a number of researchers on the Atapuerca team will tell us more.

Homo antecessor

When new fossils are found, the paleontologist compares them with those found earlier in other deposits, in an attempt to determine to which species they belong. Sometimes the comparison shows that the new fossils are unlike any others, and at this point a new species is created for them. This is the procedure which was followed with the human fossils of Gran Dolina. Following many studies and comparisons, in 1997 José María Bermúdez de Castro, Juan Luis Arsuaga, Eudald Carbonell, Antonio Rosas, Ignacio Martínez, and Marina Mosquera created the species *Homo antecessor* (*antecessor* – "pioneer," the one who comes before others).

Establishing the position of the new species in human evolution is another matter. The Gran Dolina fossils show primitive traits in dentition and other parts of the skeleton, which is to be expected given that they are 800,000 years old. These archaic characters are no longer found in later European fossils, and this is why the humans of Gran Dolina are considered to be of a different species from fossils such as the Mauer mandible, which is approximately 500,000 years old. On the other hand, the species represented in Gran Dolina is not *Homo erectus*, since it lacks the specializations of the latter species. Finally, these first European humans could represent a late population of *Homo ergaster*. However, there are a number of indications that this is not the case.

An adolescent jawbone fragment from Gran Dolina is less robust than those of *Homo ergaster*. The canine and the third molar are also reduced. Furthermore, we have the morphology of a child who died at the age of 11, which revealed something surprising. Part of the frontal bone of this child is preserved, mainly the right-hand half, which has a well-developed brow ridge. There is no doubt that in an adult the ridge would be very robust. Some of the transverse diameters of the front part of the skull have been estimated; these suggest that the Gran Dolina child had a brain larger than that of *Homo ergaster* (the brain grows very little after the age of 11). In the three best-preserved skulls of *Homo ergaster* (ER 3833, ER 3733, and WT 15000) the cranial capacity is respectively 804 cc, 850 cc, and 900 cc, while that of the Gran Dolina Child is at least 1,000 cc.

The face of the Gran Dolina Child is astonishingly modern. In *Homo habilis, Homo ergaster* and, as far as we know, in *Homo erectus*, the skeleton of the face is still fairly flat. Our face, however, is more sculpted, because the nasal opening is forward of the rest of the face, and the bones of the cheeks (the maxilla and zygomatic bone) are hollowed below the cheekbones, which thus project markedly. It is this combination of a primitive frontal bone with a modern face that makes the Gran Dolina Child not just one more fossil, but a very important specimen in our search for information about our origins.

It was always thought that the modern face was a recent development in human evolution, appearing with our species; suddenly we find that it already existed 800,000 years ago. Where do we find fossils with a modern face, of an age in between these two? We have the answer in Gran Dolina itself, where fragments of the facial skeleton of adult individuals have also been found, these showing a less sculpted form. We now know that as the individual grew up, the face grew to become very large and robust, and also more rounded as the maxillary sinuses expanded, eventually hiding the features of the child in the adult's face.

Hundreds of thousands of years later, our direct ancestors underwent an enlargement of the brain which altered the structure of the neurocranium, and a reduction in the masticatory apparatus, affecting the face, the mandible, and the teeth. These are the two cranial traits which distinguish us. The enlargement of the brain involved a fairly complete reorganization of the neurocranium together with a marked change in its shape, but the reduction in the masticatory apparatus was achieved in the simplest manner possible: the facial skeleton stops developing at an earlier point, retaining a childish appearance. In other words, our adult face is like those of the children of our ancestors.

So we see that the Gran Dolina fossils occupy an evolutionary position intermediate between *Homo ergaster* and ourselves, the only modern humans. *Homo antecessor* precedes our species but, as we shall see, it also precedes the Neanderthals (Figure 12.2), another human species different from ourselves which became extinct a few thousand years ago (practically yesterday if we think in the vast terms of geological time, and even in terms of the short duration of human evolution).

In principle it is assumed that the first humans reached the Iberian peninsula by exclusively terrestrial means, coming from Asia and crossing the whole of Europe. There is no reason to believe that the Straits of Gibraltar closed at any point during the last 3 million years, although this may well have occurred during a brief interval at the end of the Miocene, between 6.5 and 5 million years ago (i.e., too early for humans, who did

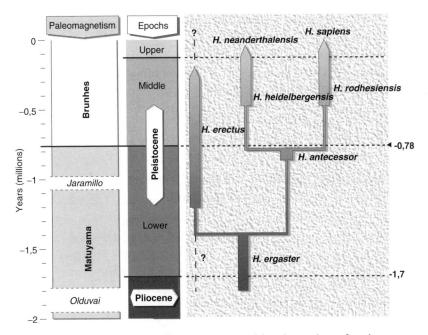

Figure 12.2 Evolutionary diagram proposed by the authors for the genus *Homo*, from the initial settlement of Eurasia. The recently created species *Homo antecessor* is included

not yet exist, to cross over from Africa). As, moreover, the currents in the Straits of Gibraltar are unfavorable for crossing, and the first humans are not thought to have had seafaring skills, there are no solid arguments for a western route, directly from Africa, for the colonization of Europe. We shall see later that the oldest known human seafaring activity occurred a few thousand years ago, and made possible the settlement of Australia and Papua New Guinea. And those who achieved it were humans of our own species (when sea level falls during the ice ages, Java and England can be reached over land, but not Australia).

But if the European populations of *Homo antecessor* came from Asia, and the Asian populations from Africa, where are the fossils of this species outside of Europe? The answer is that they have not yet been found, partly because there are no good African fossils of the same age, and the Asian fossils which might be contemporary with *Homo antecessor* are those of *Homo erectus* from the Far East. Three mandibles and a parietal bone were found in North Africa, in the Tighenif (previously Ternifine) deposit in Algeria; these are dated at 700,000–600,000 years old, later than the *Homo antecessor* fossils. There are other mandibles of similar or slightly more recent age in East Africa and Morocco. Unfortunately in Gran Dolina we have only a fragment of an adolescent mandible to compare with the African specimens.

We will therefore have to wait a little longer to get to know the African relations of the Gran Dolina fossils. From that moment on, the European branch of *Homo antecessor*, represented by the humans of Gran Dolina, and the African branch, whose fossils have not yet been discovered, followed different evolutionary histories.

Human Evolution in Europe during the Middle Pleistocene

Until recently, European fossil deposits were associated with one or other of the cold or warm periods of the Alpine glaciations, identified on the basis of the river terraces along the upper reaches of the Danube River. However, as we have already noted in the chapter on climate, *paleotemperature curves*, established on the basis of the ratio of heavy and light isotopes of oxygen in the calcareous shells of microfossils obtained from deep-sea surveys, are now used. This record is much more consistent than the continental record, and reflects changes of temperature at a global level, since the ratio of oxygen isotopes is related to the size of the polar ice caps and the corresponding rises and falls in sea level. Paleotemperature curves are divided into a series of isotopic stages, numbered from the present backward. The present day is isotopic stage number 1. All the odd-numbered stages are warm, and the even-numbered stages are cold.

The Middle Pleistocene is a division of the Pleistocene which runs from the Matuyama/Brunhes change in magnetic polarity 780,000 years ago to the beginning of isotopic stage 5. This warm stage began around 127,000 years ago, at a point when the climate was particularly warm (perhaps even warmer than now). The Middle Pleistocene is a very important period in human evolution because it was this period that saw the development of the two human species we know best – Neanderthals and ourselves.

The oldest human fossils of the European Middle Pleistocene, apart from those of Mauer and Boxgrove, are those of Arago (France) and Ceprano (Italy), all dated at more than 415,000 years old (before isotopic stage 11). The fossils found in the Arago cave include two mandibles, a facial skeleton and a right parietal bone which belong to the same individual, a coxal bone, and some other human remains. The incomplete braincase from Ceprano was discovered in a ditch dug during construction of a freeway. Morphologically it has been described as very archaic, close to *Homo erectus*. This is a very interesting specimen, but there are still many questions relating to both its chronology and its phylogenetic (evolutionary) position.

The Middle Pleistocene human fossils of intermediate age, situated chronologically between isotopic stages 11 and 8 (from 415,000 years ago to 245,000 years ago), form another group. These include the cranial remains from Bilzingsleben, Steinheim, and Reilingen (Germany), Swanscombe (England), Petralona (Greece), Vértesszöllös (Hungary), and Sima de los Huesos (another deposit in the Atapuerca Mountains in Spain, which we describe in detail below).

The most recent European Middle Pleistocene fossils are those of warm stage 7 (from 245,000 to 190,000 years ago) and cold stage 6 (between 190,000 and 127,000 years ago); this group includes, among others, the remains from Ehringsdorf (Germany) and Biache-Saint-Vaast and La Chaise-Abri Suard (France). In 1993 a complete skeleton covered with calcareous concretion was found in the bottom of a pothole in the Lamalunga cave (Altamura, Italy). This may be a late Middle Pleistocene fossil.

The teeth from Pinilla del Valle (Madrid, Spain) are also from the late Middle Pleistocene. At other Spanish sites, the Bolomor molar (Valencia), the Lezetxiki humerus (Guipúzcoa), the humerus and coxal bones from Tossal de la Font de Vilafamés (Castellón), and the mandible and other human remains from Valdegoba (Burgos) may also be of this age. In Bañolas (Gerona, Spain) a mandible with very unusual morphology (perhaps related to the strange wear shown by the teeth) was found; this is difficult to ascribe to any group. Some authors have attributed it to the late Middle Pleistocene, although the travertine limestone in which it was found is much more recent (around 45,000 years old). Finally, human fossils have been found in the Cabezo Gordo deposit (Murcia, Spain); some of these appear to be from the Middle Pleistocene.

The Pit of Bones

We have already noted that the discovery of a human fossil in a European deposit of the Lower or Middle Pleistocene is something close to a miracle. This is why these finds are so rare and so valuable. One of these extremely rare and lucky chances (for us) occurred in Gran Dolina, and the cause was an episode of cannibalism which took place about 800,000 years ago. But another site, also in the Atapuerca Mountains, holds the largest deposit of human fossils ever found, and it owes its existence to another extraordinary cause.

Not far from the trench of the abandoned railroad and from Gran Dolina, there is an extensive system of underground caverns with two entrances: Cueva Mayor (Mayor Cave) and Cueva del Silo (Silo Cave). Following a rough path for 0.5 km from the mouth of Cueva Mayor, we find a vertical shaft 14 m deep, which then continues for a further few meters on a slope and ends in a small chamber. This cul-de-sac is known as La Sima de los Huesos (The Pit of Bones), and indeed it contains a large deposit of fossil bones, embedded in clay. The bones are all of carnivores or humans: not a single herbivore fossil has been found, nor any stone tools. Among the carnivores, the majority are bears of the species *Ursus deningeri*, the ancestor of the enormous cave bear which lived thousands of years later. There are probably over 200 of these bears accumulated in La Sima. There are also remains of some lions, wolves, lynx, cats, foxes, and Mustelidae of the marten and weasel type. In terms of human remains, there are at least 32 individuals in the deposit. In both the carnivores and the humans all parts of the skeleton are represented, proving that what accumulated in this pit were complete cadavers rather than isolated bones.

The first human fossil in the Atapuerca Mountains was found here in 1976 by Trino Torres, an expert on fossil bones, and Edelweiss, a Burgos potholing group. This momentous discovery prompted Emiliano Aguirre, who was at that time supervising Trino Torres' doctoral thesis, to launch an ambitious excavation project throughout the Atapuerca Mountains. Unsupervised activities by visitors to La Sima before 1976 had altered the upper sediments of part of the deposit, so that in the 1976 excavation, the sampling carried out in 1983, and the campaigns of 1984 and 1988 work was almost exclusively on sediment that had been disturbed. Removal of this material was effectively completed by 1989, and since then the intact sediment has been excavated in a series of annual campaigns.

The work to remove the disturbed sediment (in backpacks until 1987), extract several tons of large limestone blocks, and adjust the infrastructure of the cave, was a major undertaking. Moreover, the sediments disturbed by visitors contained almost exclusively bear bones, with very few, in any case very fragmented, human remains. But all of this work was rewarded in 1992 with the discovery of three very complete skulls, including Skull 5, at this moment in time the best-preserved skull in the fossil record of human evolution (Figure 12.3). Excavations in La Sima are still continuing to produce a large quantity of very well-preserved human fossils.

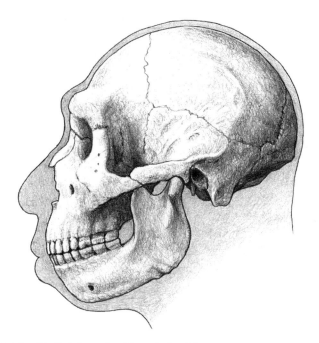

Figure 12.3   Skull 5 from La Sima de los Huesos

An extensive program to date the speleothems (mineral deposits) in La Sima, as well as dating the human and bear remains directly, using the uranium series and ESR methods, is underway in order to establish the age of the human fossils. This work is being carried out by geochronologists James Bischoff and Christophe Falguères. The results obtained, which are in keeping with the fauna, give an age of about 300,000 years for the human fossils.

How did this deposit form? The cavities close to La Sima, within the Cueva Mayor, also contain abundant remains of bears of the same species as La Sima, which no doubt used these chambers to hibernate, and sometimes died there from natural causes (sickness, age, or simply because they had not accumulated enough fat in the autumn to last them through the long winter). Some of the bears in the cave might have fallen accidentally into La Sima, from which they would not have been able to escape. The odd lion, wolf, or other creature, attracted by the odor of carrion, might occasionally have jumped into this natural trap (there are other deposits of this type at the base of potholes). At that time they would not have had to make the long journey currently taken from the mouth of the Cueva Mayor, since there was a much more direct entrance, which later fell in.

However, the presence of at least 32 human cadavers in the deposit is more difficult to explain than that of the bears. The absence of herbivores and stone tools means that we can eliminate the possibility that humans carried on their habitual activities here (even if there was at the time a less rugged access route, apart from the pit of La Sima itself). Nor does it seem reasonable to attribute the accumulation of human cadavers to the action of carnivores, again because of the absence of herbivores in the deposit. Two alternatives remain. One is very vague: some kind of unidentified catastrophe. The other, which we prefer, is that other human beings, who could have dropped the bodies of their dead companions into a pit in a hidden, dark part of the cave, were responsible for the accumulation. This would be the oldest known evidence of a funerary practice.

The paleodemography (distribution of ages of death) of the fossils from La Sima has been studied by José María Bermúdez de Castro, as part of his work in dental paleoanthropology. From this we know that most of the individuals whose bodies accumulated in the deposit were adolescents aged between 13 and 19, and young adults, less than 30 years old. This age distribution is one more mystery among the many surrounding La Sima. Why are there no children, or mature and old adults? Did all the individuals die within a short period (months or years), or is the accumulation the result of a practice which lasted for generations? Is it possible that everyone at that time died very young, before the age of 40? This latter hypothesis is difficult to entertain, since we know that the period of maturation in this fossil population was similar to our own, and we therefore have to conclude that their potential life expectancy was roughly the same as ours today. Even chimpanzees, whose period of maturation is much shorter than that of humans, live beyond the age of 40 in the wild. This is, in any case, just one of the problems currently being investigated; it is also a matter of great interest in relation to finding out why so many cadavers accumulated in La Sima de los Huesos.

The present authors, together with José Miguel Carretero, Ana Gracia, and Carlos Lorenzo, have spent many years studying the human fossils of La Sima; among their areas of interest is the difference between the sexes in terms of body size. Many experts maintain that since sexual dimorphism in body volume was much greater in the first hominids than it is now, there must have been a more or less constant gradual reduction over the course of human evolution. In this case, the Middle Pleistocene fossils, including the Neanderthals of the Upper Pleistocene, should still show greater sexual dimorphism than we do.

Two serious problems arise when we try to verify this hypothesis. One is the proverbial scarcity of fossils, particularly of the postcranial skeleton, which is used to estimate body size. Thus researchers are forced to combine fossils from deposits in different places and eras in order to put together a sample, which is in any case very small. The problem is not so great in the case of the Neanderthals, for which there is a much larger (though also dispersed) sample.

The second problem is methodological. In order to compare women and men we first need to establish the sex of the fossils. The criterion of size is generally the one used: the large bones are assumed to be those of males, and the smaller bones to belong to females. But when the difference in size is not enormous, as it is in gorillas and orangutans, there is a wide band of overlap of medium-sized and small male individuals and large and medium-sized females, in which it is difficult to establish the sex of the remains.

In La Sima de los Huesos we have a broad sample of individuals from a single biological population. Moreover, our approach eliminates the methodological problem, because what we are studying is the variation in the sample, without assigning sex *a priori* to the fossils; we start from the premise that the greater the difference between the sexes in the population, the greater will be the variation observed in the sample.

In addition to measuring the bones of the postcranial skeleton, which reflect differences in body weight, we also analyzed variation in brain size.

In La Sima there are three skulls whose cranial capacity is known. One is Skull 5, with a cranial capacity of 1,125 cc; the second is Skull 4, a complete calvarium with a cranial capacity of 1,390 cc. The third is Skull 6, fairly complete, and corresponding to an adolescent aged about 14, which has a cranial capacity of about 1,220 cc.

The result we have obtained is that the sexual dimorphism in body and brain size in the population represented in La Sima (from about 300,000 years ago) was no greater than our own.

As La Sima de los Huesos is not a place in which humans would have lived, we have no information on their behavior here. However, in one of the Trinchera del Ferrocarril (Railroad Trench) deposits of the Atapuerca Mountains, known as La Galería (the Gallery), 13 archeological levels have been excavated which give us some idea of human activity and technology of the same period as La Sima and a little after. Two human fossils, a jawbone fragment and a skull fragment, have also been found in La Galería.

13

The Neanderthals

"The Red Flower?" said Mowgli. "That grows outside their huts in the twilight. I will get some."

Rudyard Kipling, *The Jungle Book*

The Way They Were

The discovery of the skull of a Neanderthal child aged two or three in Engis (Belgium) in 1830 ought to have marked the beginning of paleo-anthropology. But, like the Neanderthal skull excavated in the Forbes quarry in Gibraltar in 1848, its true significance in human evolution was not recognized at the time. Much more famous, and also controversial, was the discovery in 1856 of a skeleton – which gives its name to the entire group – in the Feldhofer cave near Düsseldorf, in the Neander valley (*Neander Thal* in the old German spelling, Neander Tal in modern-day spelling).

In 1859 Charles Darwin published *On the Origin of Species*, inching open the door to the search for fossil antecedents of our species. A few years later, in 1871, Darwin opened wide this door with his book *The Descent of Man, and Selection in Relation to Sex*. Nevertheless, the status of Neanderthals as members of an extinct human form distinct from our own was not definitively recognized until the discovery of further fossils, in particular those found in Spy (Belgium) in 1886, made it impossible to continue considering them as atypical or pathological specimens of modern humans.

Many more hominid fossils, some of them of incomparably greater antiquity, have been found since, but these have failed to usurp the place occupied in the popular imagination by the Neanderthals, and they continue to be seen as the quintessential fossil humans. Nevertheless, the Neanderthals were not the brutal, apelike beings, incapable of walking erect, that many people suppose. They were strong and at the same time skillful gatherers of plant products, hunters, and carrion eaters. They had a wide variety of highly refined stone tools. They used fire systematically, looked after their old and disabled, and buried their dead.

We know them well, almost as if they were still living, because in fact Neanderthals are the human fossil type which has been studied in the greatest detail, and of which we have the most remains. The Neanderthals were not very tall, with the males averaging 170 cm in height, and the females 160 cm. However, they had great physical strength. Christopher Ruff, Erik Trinkaus, and Trenton Holliday have studied the body weight of humans over the last 2 million years, and have concluded that the Neanderthals were the humans with the greatest muscle mass – in other words, the strongest (although preliminary studies, yet to be published, indicate that the humans of La Sima de los Huesos were even stronger). On average (for both sexes) Neanderthals weighed 70 kg, compared to the average for our species (taken as a whole over the different living populations and both sexes) of around 58 kg – 24 percent less.

The La Ferassie 1 fossil, from France, which is probably male, gives a good example of the strength of the Neanderthals: it has an estimated height of 171 cm, and a calculated weight of 85 kg. Moreover, John Kappelman believes, in our view quite reasonably, that the weight estimates made by Ruff and his colleagues are on the low side, because they use formulas based on the relationship between body weight and bone size in normal modern subjects. However, the development of the insertions of the muscles in the bones indicate that the musculature was much more developed in Neanderthals than in normal modern humans, and it might have been more appropriate to make the comparison with elite athletes, such as weightlifters, or javelin or discus throwers. As a result, it is entirely credible that La Ferassie 1 weighed as much as 90 kg of pure muscle, despite the individual being only 171 cm tall (Figure 13.1).

In the German deposit of Schöningen, a number of wooden spears around 400,000 years old have almost miraculously been preserved in the peat of an ancient bog where humans hunted horses. The longest is 2 m long and was, according to Hartmut Thieme, who discovered it, designed to be thrown. We need only imagine a group of muscular Neanderthal hunters armed with spears like these to see that they were hardly defenseless creatures.

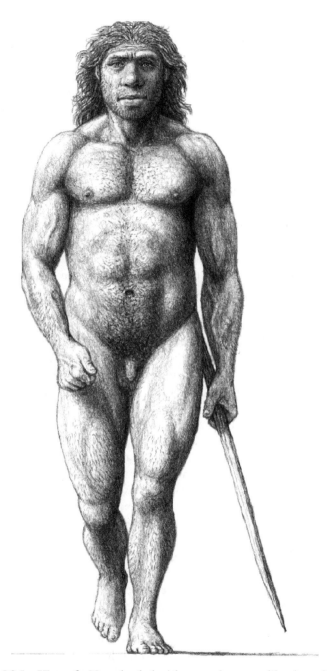

Figure 13.1   View of a Neanderthal with a wooden spear like those found in Schöningen

Many authors believe that this strapping physique corresponds to a well-known law of biogeography known as Bergmann's rule, according to which the populations of a warm-blooded species living in cold regions tend to have a larger body than those living in hot regions. In this way, the body shape approaches that of a sphere, the form which has the smallest surface area in relation to volume (the musk ox is a perfect example of an animal from a polar environment). This relative reduction in surface area reduces heat loss from the body through radiation.

In addition, the Neanderthals had relatively shorter forearms and lower legs (below the knees). However, populations living in hot regions follow the converse Allen's rule, their long, thin arms and legs maximizing the surface area to volume ratio (the dromedary is a good example of this biotype). Within our species these laws are also observed, as can easily be verified if we compare an Inuit from the Arctic Circle with a Tutsi or a Tuareg, living in the hottest regions of Africa. In fact, the proportions of the Turkana Boy (as the reader will remember, a *Homo ergaster* from 1.5 million years ago) would, according to his discoverers, be similar to those of the human populations living in this region today (see Figure 7.5). The first colonizers of Eurasia, who came from Africa, must also have been tall and slim (a hypothesis which further discoveries of new fossils in Gran Dolina will enable us to confirm); hundreds of thousands of years later the Neanderthal physique reflected an adaptation to the rigors of a cold climate, which enveloped much of the European continent during the ice ages.

One notable characteristic of the Neanderthals is a very elongated and flattened pubic bone (specifically, the upper, horizontal ramus of the pubic bone). The elongation of the pubis is probably not a trait exclusive to the Neanderthals, since it is also found in australopithecines, as we have seen. Before any complete Neanderthal pelvises had been found, and only fragments of pubic bones and other broken pieces were available, some authors believed that, given the extraordinary length of the pubis, the birth canal would also be very large. There were even some who suggested that pregnancy would have been a month or so longer than in our species, so that Neanderthal babies came into the world more fully developed. However, the discovery of a very complete Neanderthal pelvis in the Kebara deposit in Israel showed that although the pelvic structure was different from ours, the birth canal was not substantially larger – perhaps a little larger, but since the adult brain was also a little larger than ours, the stage of development of the newborn would have been comparable with that of our babies. For new information on this question we will have to await publication of the results of studies currently under way on the complete pelvis recently discovered in the Sima de los Huesos deposit in Atapuerca, Spain.

It may have come as a surprise to the reader that Neanderthals' brains were not smaller, but larger than ours: the average capacity was 1,500 cc, whereas the modern average, calculated for all populations, is around 1,350 cc. However, given that because of their great muscle mass the body weight of the Neanderthals was higher, it is likely that the index of encephalization (the only way we have in paleontology to express, in terms of measurement, something like intelligence) was slightly lower than our own.

What is certain is that Neanderthals and ourselves are the two most highly encephalized human forms in history. However, this increase in brain size occurred independently in the two cases. While the modern neurocranium is high, with a high forehead, that of the Neanderthals was very elongated from front to back (Figure 13.2). Because of this anterio-posterior elongation of the brain cavity, the occipital bone extended backward, forming the characteristic bump known as the "occipital bun" of Neanderthal skulls, which is clearly visible from the side. The paleoanthropologist Giorgio Manzi has shown how the increase in brain size in the Neanderthals did not involve a substantial change in the archaic structure of the skull, which remained low, with a flat forehead; it was simply modified to contain a brain which was sometimes enormous. Manzi has said that in Neanderthals the enlargement of the brain caused a conflict similar to that which would be generated if a small saloon car had to be adapted to take a Formula 1 engine.

Viewed from the back, there is an important modification in the neurocranium of the Neanderthals. In *Homo ergaster* and *Homo erectus* the skull was widest at the base, and the two sides converged toward the top of the cranial vault, like a house with the walls leaning inward. In Neanderthals and modern humans the widest part of the neurocranium is in the middle, above the parietal bones; the skull is narrower at the base. However, while our skull, viewed from the back, looks like a house with the walls leaning outward from the base toward the roof (i.e., inclined outward), that of the Neanderthals is rounded.

Neanderthals show other neurocranial traits which are exclusive to them, some of them not very noticeable but extremely useful for tracing the origins of these humans. For example, above a not very marked occipital ridge, sunken in the center, Neanderthals have a depression technically known as the *suprainiac fossa*. To cite just one other character, in Neanderthals the mastoid process (the bony projection of the temporal bone, where part of the sternocleidomastoid muscle originates) projects hardly at all beyond the base of the skull (Figure 13.2).

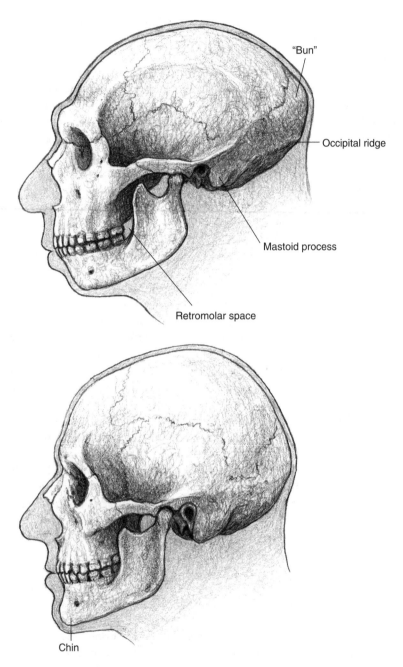

Figure 13.2   Neanderthal skull (top) and modern human skull (below)

Neanderthals have an easily recognizable brow ridge: it forms a regular curve above the eye sockets like an arc of a circle, it is rounded in cross-section and it continues between the eyes. Moreover, in general it is completely hollowed out inside by highly developed frontal sinuses.

Finally, the Neanderthal face is unique among the hominids (Figure 13.3). In other fossils, and in ourselves, the bones below the eye sockets and either side of the nasal opening (the maxilla and the zygomatic bone) form a bony surface which faces forward. We have already seen how in our species the lower part of this surface is hollowed, allowing the cheekbones to project above. In Neanderthals, in contrast, this surface faces diagonally, giving them a wedge-shaped face. It is as if the nasal opening had moved forward, stretching the sides of the face (this typical morphology of the Neanderthal face is technically known as medio-facial prognathism). In the foremost part of this pointed face there is a very wide nasal opening.

Thus Neanderthals have a large nasal cavity (with its roof formed by nasal bones which are almost horizontal, as a result of the forward shift of the nasal opening), which some authors have interpreted as an adaptation to a glacial, cold, dry climate. The air would be warmed and humidified in this cavity before passing into the lungs. Moreover, the highly developed frontal sinuses (above the nasal opening and over the eye sockets) and maxillary sinuses (either side of the nasal cavity) would contribute to form an air chamber which would insulate the brain, an organ which is very sensitive to changes in temperature. An enlarged face, like a great hollow mask, would thus be interposed between the brain and the cold outside.

There are authors, however, who interpret the facial morphology of the Neanderthals differently, in biomechanical terms (although the two explanations are in fact compatible). The front teeth of the Neanderthals show very rapid wear, indicating intense use, as if the mouth was often used as a "third hand" to hold and pull on objects. The pointed shape of the face, it is suggested, would serve to divert the force generated in the bone by this activity toward the sides.

Another characteristic which is almost exclusively Neanderthal is that in the jawbone the teeth are shifted forward in relation to the bone, to the extent that there would be room for a fourth molar. This empty space, known as the retromolar space, means that when seen from the side the Neanderthal jawbone has a hollow between the last molar and the front edge of the mandibular ramus (the branch of the jawbone which runs upward toward the joint with the base of the skull).

In the living Neanderthal, this face would have been distinguished by very wide, prominent nostrils, the absence of cheekbones, a sloping

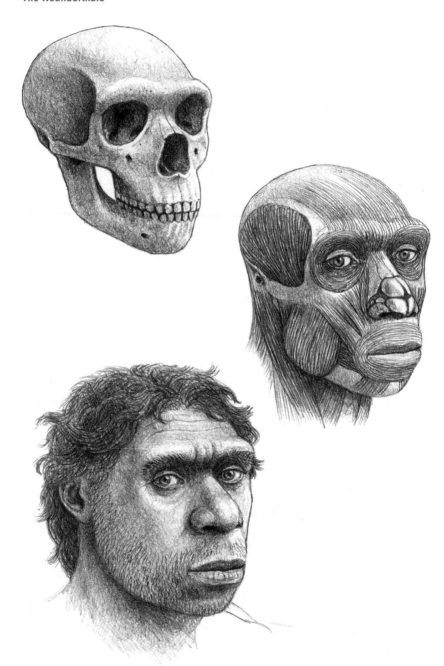

Figure 13.3   Reconstruction of the head of a Neanderthal

forehead, strongly overhanging brows, and a jaw with no chin (Figure 13.3); a well-developed chin is a trait exclusive to ourselves (Figure 13.2).

Finally, the Neanderthals' bones were thick, although this is a characteristic found in all humans since *Homo ergaster*, except in our species, where the walls of the bones have become markedly thinner. But while in *Homo ergaster* the bones of the head were the thickest, in Neanderthals the thickest bones were those of the body; the medullary cavities, which are found inside the long bones like the femur, humerus, and tibia, are very narrow as a result of the thickening of the bone walls. The reason for the production of so much bone is not actually known, but it is certain that the Neanderthals needed a great deal of calcium. This is not found in meat: in fact an excess of animal proteins acidifies the blood, requiring neutralization by the release of calcium from the skeleton and thus depleting the mineral levels in the bones (carnivores have bones with thin walls, while in herbivores they are thicker). One source of calcium is milk and dairy products, which Neanderthals certainly did not eat. Their only source of calcium, therefore, was plants; Neanderthals no doubt had to consume these in large quantities.

Life and Death among the Neanderthals

The Neanderthals' stone tool industry is known as Mousterian, and is classified within Mode 3 or the Middle Paleolithic (Figure 13.4). The characteristic of this technical mode is that the cores are worked by giving them a specific shape (similar to a turtle shell), in order to obtain flakes from them, which will then be retouched to achieve the final form. This sequence of operations is known as the Levallois technique, and various tools are obtained from each core, thus making better use of raw material and effort. Moreover, it is clear that it implies a distinct capacity for abstraction, since the stone is not worked directly to produce the tool; instead an intermediate step (the Levallois core) is introduced.

The Mousterian extended throughout Europe, the Near East, and North Africa, around the Mediterranean, while other Mode 3 or Middle Paleolithic industries are found in the rest of Africa (where they are grouped together under the term "Middle Stone Age"). Mode 3 originated between 300,000 and 200,000 years ago, depending on the region, appearing first in sub-Saharan Africa and later in Europe. Once more we come up against the problem of how the technique arrived in Europe. Was it brought by people who came from Africa? And if so, what was their

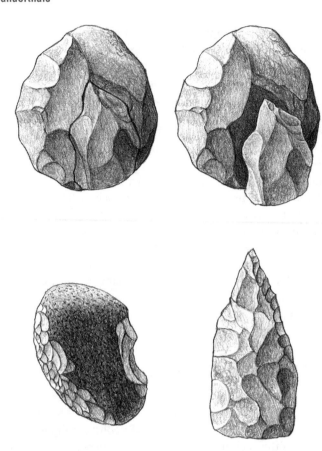

Figure 13.4   Some representative Mode 3 tools. *Top*: Levallois core and point extracted from it; *below left*: retouched scraper of the la Quina type; *below right*: bifacial point (Middle Stone Age)

relationship with the local populations, those responsible for the Acheulian (Mode 2) industry? Or was it only the new technique that traveled, without population movements? We shall return to this problem later.

Two aspects of Neanderthal behavior which bear striking similarity to our own are the *use of fire* and the practice of *burying the dead*. There are animals which, albeit in very simple ways, select and even modify natural objects to use them as tools. However, no animal species other than ours knows the technology of fire, nor does any bury its dead or conduct any ceremony for them. The practice of burial is thus a "humanizing" trait. It is difficult now to imagine human life in the wild without the use of fire. In fact, what is surprising is that there have been humans who did not have fire, at least in Europe and parts of Asia (in regions far from the equator). In many senses fire makes us human, and the absence of fire places us on the same level as animals.

It is not known for certain when the capacity to make and control fire first appeared among humans. It is possible, although not certain, that it was used in the Middle Pleistocene in places like Zhoukoudian in China, Terra Amata in France, Vértesszöllös in Hungary, La Cotte de St. Brelade (on the island of Jersey, Great Britain), and Bilzingsleben in Germany – all deposits which are 200,000 years old or more. However, the generalized and systematic use of the technology of fire, with all it implies in terms of protection, heat, light, and so on, emerged somewhat less than 200,000 years ago. From this time on there are well-structured fireplaces in the deposits, which leave no room for doubt that fire had been mastered. At this point the humans living in Europe, those who made the fires, were Neanderthals, now masters of the "Red Flower."

It has always seemed evident that the Neanderthals buried their dead, and many of the Neanderthal skeletons excavated in cave deposits have historically been considered the result of this burial practice (for example, eight Neanderthals were found entombed in the French deposit of La Ferassie). It is less certain whether the burials were accompanied by a ritual, a ceremony with some symbolic significance, although in some cases it has been suggested that there is evidence of this. For example, the earth covering the skeleton of one of the Neanderthals from Shanidar (Iraq) contained grains of pollen from flowers which were thought to have been placed on the body; several pairs of mountain-goat horns found around the skeleton of a nine-year-old child in Teshik Tash cave (Uzbekistan) are thought to have been placed there deliberately; on the skeleton of a two-year-old child in Dederiyeh (Syria) a triangular flint tool was found at the level of the heart and a limestone "tombstone" next to the head; a deer jawbone was found on the pelvis of a ten-month-old child at Amud (Israel); the Le Moustier adolescent, in France, was thought to have been sprinkled with ochre and buried in a flexed position, accompanied by offerings.

However, in their book *In Search of the Neanderthals*, paleoanthropologist Chris Stringer and archeologist Clive Gamble very reasonably call into question this evidence of ritual, which of course is open to other, more prosaic interpretations. A casual association of human bones with animal remains or tools, which produces the appearance of something intentional, is always possible in a deposit where there is an abundance of animals and stone tools (and pollen is brought in by the wind). Moreover, some of the excavations, such as that of Le Moustier, are old, and there are reasons to question how strictly they were conducted (each generation of scientists criticizes the methodology of previous generations, although not always with justification). Stringer and Gamble extend their skepticism even to the suggestion that Neanderthals buried their dead.

We believe that they did do so, because the deaths of Neanderthals whose skeletons have been found in caves cannot all be due to natural causes. Furthermore, there is evidence that the Atapuerca humans were already conducting funerary practices (though not burials exactly) 300,000 years ago, in accumulating the bodies of their dead in La Sima de los Huesos. It is true that this is a unique case, but it is also possible that burials or other funerary practices were conducted at this time outside of caves, leaving us no fossil record or knowledge of them.

The Beginning and End of the Neanderthals

The Neanderthals properly speaking, i.e., with all or the majority of their typical characteristics, existed in Europe some 230,000 years ago. The fossils from Ehringsdorf, Biache-Saint-Vaast, la Chaise-Abri Suard, and others from the end of the Middle Pleistocene can already be considered Neanderthal.

The key to understanding the evolutionary position of the group of Middle Pleistocene fossils of intermediate age (between 415,000 and 245,000 years old), such as those of Steinheim, Swanscombe, Reilingen, Vértesszöllös, Petralona, and La Sima de los Huesos, is in the broad sample from the Spanish deposit. Although in general their morphology is primitive, the fossils from La Sima de los Huesos show incipient Neanderthal traits in the occipital bone and the face, and other more clearly Neanderthal features in the mandible, for example the presence of a retromolar space (see Figure 12.3). The postcranial skeleton also bears traits in common with the Neanderthals, such as the morphology of the pubis and humerus, among others. But there are so few remains of these bones in the fossil record outside of La Sima de los Huesos that at present it is difficult to know at what point in human evolution these traits appeared (it may even have been before the colonization of Europe).

The Steinheim skull may possibly have been similar to those of La Sima de los Huesos, though it is difficult to tell because it is incomplete and badly deformed. More fragmented fossils, such as those from Swanscombe, show a more typically Neanderthal occipital morphology (with a ridge that is depressed in the center and a broad suprainiac fossa). The

Petralona skull has a more Neanderthal face than those from La Sima de los Huesos, but its occipital bone is less Neanderthal. Taken together, this group of fossils indicate that the Neanderthals evolved in Europe over hundreds of thousands of years in geographical and genetic isolation. In this sense the Neanderthals are the original Europeans, a native, local human species – what is known in biology as an *endemic* species.

The Neanderthals' roots extend far back into the past, but how far back do they go? Some authors, such as Chris Stringer, group the Mauer mandible with the more modern fossils of Arago, Bilzingsleben, and Petralona together to form a species known as *Homo heidelbergensis*, with the Mauer mandible as the type specimen.

African specimens dated at between 600,000 and 250,000 years old, such as the skulls of Bodo (Ethiopia), Ndutu and Eyasi (Tanzania), Salé (Morocco), Elandsfontein (South Africa), and Broken Hill (Zambia), are added to the European fossils of this species.

It is also suggested that *Homo heidelbergensis* is represented in Asia by the Dali skull and the Jinnishuan skeleton, both from China, and dated at between 200,000 and 300,000 years old. Precise dating of these fossils is important, because they might be contemporaneous with the last *Homo erectus* fossils in China, Skull 5 from Zhoukoudian, and an incomplete skullcap from Hexian cave.

According to this theory, the place of *Homo heidelbergensis* in human evolution would be that of the last common ancestor of Neanderthals and modern humans (Figure 13.5). The cranial capacity of the species ranges from 1,000 to 1,400 cc, the lower value in the range approximating to the average for *Homo erectus*, and the upper to the average for Neanderthals and modern humans.

In a recent article Robert Foley and Marta Lahr elaborate an evolutionary model which they call the Mode 3 hypothesis. According to them, the species which invented Mode 3, between 300,000 and 250,000 years ago, was a species which evolved in Africa from *Homo heidelbergensis*. They suggest that this new species (whose Latin name we shall omit, so as not to add further to the terminological confusion) had a larger brain and was more intelligent than previous species, as demonstrated by their advanced (Mode 3) technology. Some members of the Mode 3 Species migrated to Europe around 250,000 years ago, in time replacing the European population of *Homo heidelbergensis* and evolving into the Neanderthals. In Africa the Mode 3 Species evolved to create our species, *Homo sapiens*.

However, in our view all these authors are mistaken. Our analysis of morphological traits leads us to conclude that the European populations such as that of Mauer, living around 500,000 years ago, were already on

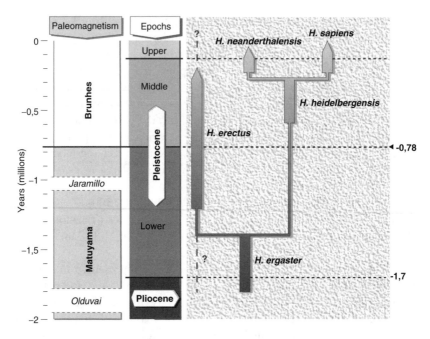

Figure 13.5 Alternative evolutionary diagram to that of the authors, in which *Homo heidelbergensis* appears as common ancestor of the Neanderthals and modern humans

the evolutionary line of the Neanderthals. The rules for creation of species state that the name of the species should be related to its type specimen; as the type specimen for *Homo heidelbergensis* is the Mauer mandible, this species should actually be exclusively European and the ancestor of Neanderthals. The last common ancestor of Neanderthals and modern humans is *Homo antecessor*, created on the basis of the Gran Dolina fossils, which are more than 780,000 years old (see Figure 12.2). As we shall see later, there are data from molecular biology which support our position.

The species *Homo heidelbergensis*, as we understand it, would cover fossils from the Mauer mandible to the fossils of La Sima de los Huesos and all those in which primitive traits predominate, although they may show some incipient characters indicating that they are ancestors of the Neanderthals. The fossils of the end of the Middle Pleistocene (from about 230,000 years ago onward), on the other hand, can to all intents and purposes be considered true and complete Neanderthals.

The two skulls from Saccopastore (on the outskirts of Rome) and the set of fossils from Krapina (Croatia) are from the beginning of the Upper Pleistocene (the latter have been reliably radiometrically dated, but the date for the Saccopastore fossils is less certain).

From this time on Neanderthals appear in abundance in European deposits (relative, of course, to the general scarcity of the fossil record), and they are also found in Central Asia and the Middle East, regions to which they migrated from Europe. Fossils as emblematic as those of Le Moustier (which gives its name to the Mousterian), Guattari 1 (Monte Circeo), and La Chapelle-aux-Saints lived in Europe less than 60,000 years ago. These Neanderthals show some new characteristics with respect to earlier specimens, and are often called "classic Neanderthals." The description of a Neanderthal that we gave earlier corresponds most particularly to these classic forms. As we shall see, there were still Neanderthals in Europe 30,000 years ago, perhaps even more recently, before their trail disappeared forever.

The Spanish record of the Upper Pleistocene is rich in Neanderthals and includes fossils from the deposits of Agut (Barcelona), Axlor (Vizcaya), Cova Negra (Valencia), Gibraltar (Devil's Tower and Forbes Quarry), Gabasa (Huesca), La Carihuela (Granada), Los Casares (Guadalajara), Mollet I (Gerona), and Zafarraya (Málaga). There is good reason to hope that many more Neanderthals will be discovered in the Iberian peninsula over the next few years.

14

The Origins of Modern Humanity: The Fossil Evidence

Resolve me strangers, whence, and what you are;
Your business here; and bring you peace or war?
Virgil, *Aeneid*

Neanderthals and Modern Humans

In the European deposits which contain unbroken archeological sequences, the Mousterian (Mode 3) industry is abruptly replaced by the Upper Paleolithic Aurignacian, or Mode 4, industry (Figure 14.1). Technically, Mode 4 is characterized by the preparation of elongated cores to obtain fine flakes and parallel edges, at least twice as long as they are wide. These flakes were then retouched and made into a wide variety of tools, including *burins* (beveled tools used to work bone, horn, and marble) and *end-scrapers* (flakes with one end retouched, used to prepare hides). This technique allowed for maximum use of the raw material, obtaining more blade length (if we add together the length of the blades of all the tools made from the original raw material) than any other method. Mode 4 is also distinguished by the use of bone, ivory, and horn as raw materials for tools and items of personal adornment. Art also appears in association with Mode 4 industries, in the form of portable figures of animals and people, and cave paintings and engravings (although these artistic expressions are not found with the first Mode 4 industries, appearing only some thousands of years later).

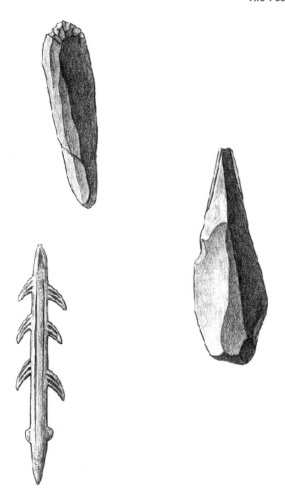

Figure 14.1   Representative Mode 4 tools. *Top*: end-scraper; *left*: bone spear; *right*: double-beveled burin

However, in some deposits on the Atlantic coast of the Iberian peninsula and in western and central France, intercalated between the last Mousterian and the first Aurignacian levels, we find intermediate levels containing an industry with characteristics common to both technical modes (3 and 4). This industry is known as the Chatelperronian; the Szeletian industry of central Europe and the Uluzzian industry of Italy may be equivalents. The Chatelperronian is an evolution of the Mousterian industry which includes elements of Mode 4, such as long flakes, and tools made from bone and ivory.

The Aurignacian spread quickly through Europe, about 40,000 years ago or a little earlier. We have two examples which have been studied in depth where the Mousterian is abruptly replaced by the Aurignacian, in the deposits of L'Arbreda and El Abric Romaní in Catalonia, Spain; here radiometric data obtained by James Bischoff indicate that the change happened around 40,000 years ago. On the Atlantic coast the oldest Aurignacian industry, found in the El Castillo cave, has been dated by Bischoff at 39,000 years old. Who made the Chatelperronian and the Aurignacian tools?

The human fossils found so far with Mousterian (Mode 3) tools in Europe are always Neanderthals, while those found with Mode 4 industries are always modern humans. Fossils associated with the Chatelperronian industry include those of the Grotte de Renne at Arcy-sur-Cure (near Auxerre, France), which consist of isolated teeth and bone fragments, and the partially preserved skeleton from Saint-Césaire (Charente-Maritime, France).

There is no doubt that the Saint-Césaire fossil is of a classic Neanderthal individual who lived around 36,000 years ago. The Chatelperronian level associated with the human remains in the Grotte de Renne at Arcy-sur-Cure is dated at 34,000 years old, and it contains bone and ivory tools, together with items of personal decoration such as perforated or grooved animal teeth and ivory beads or pendants (all decorative pieces typically associated with Mode 4). The nature of the human fossils in the Grotte de Renne was not clear because they were so incomplete. However, Fred Spoor, Jean-Jacques Hublin, and their colleagues have demonstrated that the morphology of the bony labyrinth (inner ear) of Neanderthals was different to our own. This led to the identification of the temporal bone of a child aged approximately twelve months as Neanderthal (the bony labyrinth is fully formed before birth – fortunately, since it is impossible to establish from the external morphology whether a temporal bone from such a young individual is Neanderthal or not).

Thus, in the only two deposits which contain human fossils in association with Chatelperronian industry, the fossils are of Neanderthals. On the other hand, the very recent dates of the Chatelperronian, later than the first Aurignacian industries, rule out the possibility that Neanderthals evolved into modern humans in western Europe, if we assume that modern humans were responsible for the first European Mode 4 industries.

The cave paintings, the most spectacular artistic expressions of prehistory (and perhaps of the entire history of art), also begin early: the marvelous images of animals recently discovered in the Grotte Chauvet (in Ardèche, France) have been radiocarbon dated at about 30,000 years old, and no one would venture to attribute them to Neanderthals. Other cave paintings, also recently discovered, in the Grotte Cosquer (Bouches-du-Rhône, France), are about 27,000 years old. The oldest Upper Paleolithic figurines, the stone figure of a woman from Galgenberg (Austria) and the small ivory sculpture of a horse from Vogelherd (Germany), are perhaps 32,000 years old.

However, unfortunately we do not as yet have any human fossils associated with the first Aurignacian industries in Europe. The modern skeletons from the Cro-Magnon deposit (France) are assigned to the end of the Aurignacian, between 30,000 and 28,000 years ago. An interesting fossil in this respect is the Hahnöfersand frontal bone (Germany) which, although it has no context, has been radiocarbon dated at 36,000 years old. As we understand it, the morphology of the brow ridge indicates that it should be classified as a modern human; it is a shame that its age is so problematic. In central Europe the modern skeletal series from Mladec (in Moravia, Czech Republic) is dated at about 32,000–30,000 years old, and is associated with an Aurignacian industry.

The last Neanderthals (like the Saint-Césaire specimen) show no signs of evolution toward modern humans; in fact they are absolutely "classic" Neanderthals. Similarly, the Mladec remains cannot be considered transitional forms: they are entirely modern, albeit robust, humans. This means that the possibility of modern humans in Europe originating through evolution from the local Neanderthals can also be ruled out categorically from the paleoanthropological point of view. We have to conclude that modern humans in Europe are immigrants who came from elsewhere. Study of the body proportions of the first modern humans in the European record gives us a clue to their origins: their gracile biotype resembles that of people from equatorial regions, less stocky, taller, and with longer arms and legs than the Neanderthals.

The last Mousterian industries, which are found in Portugal, Spain, and Italy, and are dated at exactly 30,000 years old, are contemporary with or even later than the modern skeletons of Mladec. Archeologists Gerardo Vega and Valentín Villaverde have pointed out that these Mousterian industries survived in the Iberian peninsula after they had disappeared from other regions of Europe. Confirming these observations, archeologist Cecilio Barroso and paleoanthropologist Jean-Jacques Hublin have found some classically Neanderthal human fossils in the Zafarraya deposit (Málaga, Spain); these include a mandible and a femur dated at 30,000 years old and associated with a Mousterian industry.

It is thus certain that the last Neanderthals were still living in southern Europe when modern humans were already painting rhinoceros, lions, and bison on the walls of the Chauvet cave, and 10,000 years after the first modern humans settled in the Iberian peninsula. However, this is not to give the impression that modern humans advanced from east to west like a steamroller, crushing all the Neanderthals in their path. The two species must have shared the map of Europe for thousands of years, forming a mosaic of intermingled populations. As we have seen, about 40,000 years ago there were already modern humans in the regions of Catalonia and Cantabria in Spain, and thousands of years later Neanderthals still survived well to the north of the Pyrenees. We may imagine that the pockets of Neanderthal populations gradually diminished until the last of them disappeared.

This long coexistence of Neanderthals and modern humans (popularly known as Cro-Magnon Man) is one of the chapters of prehistory which has most excited popular attention, giving rise to tales such as *La Guerre du Feu* (Quest for Fire, 1911) by J. H. Rosny-Aîné, made into a film by Jean-Jacques Annaud in 1981, and Jean M. Auel's saga *The Clan of the Cave Bear*. One particularly intriguing aspect of this coexistence between Neanderthals and modern humans is the origin of the Chatelperronian and other similar technologies. Had the Neanderthals developed the use of bone, horn, and ivory to make tools and decorations, and the production of long, thin stone blades, of their own accord? Did they do it independently in a number of regions of Europe? Or did they copy it from modern humans by watching them work, or examining abandoned objects? Might some of the elements found in the Grotte de Renne at Arcy-sur-Cure, such as the ivory pendants, come from an exchange between the two types of humans? As yet we have no solution to these puzzles.

Two Intelligent Human Species

From the paleontological point of view, Neanderthals are a different species (*Homo neanderthalensis*) from modern humans (*Homo sapiens*). This means that Neanderthals are the result of a long process of evolution independent of our own, originating from a common ancestor. As a result of this separate, divergent evolution, the difference between Neanderthals and modern humans is much greater than between the various modern populations (Inuit, Aboriginal Australians, Zulus, and Basques, for example).

When defining a new living species, the genetic criterion is usually used: members of a new species cannot mate with those of another species and have fertile offspring which could in their turn reproduce with individuals from the population of the father or the mother, or with one another. The new species must be genetically isolated from others. In the case of the Neanderthals, we know of no fossil hybrids of them and our ancestors; moreover, modern Europeans do not carry genes inherited from the Neanderthals. However, the fact that there was little exchange of genes does not mean that it was impossible, and this is the condition for defining a species on the basis of genetic criteria.

In paleontology, the evolutionary concept of the species proposed by George Gaylord Simpson (1902–84) makes more sense. According to this concept, a species is a continuum of populations which succeed one another in time and follow their own evolutionary trajectory, independent of other species, and continue over a considerable period: what counts is that there is genetic continuity between generations and that isolation is maintained. According to this criterion, Neanderthals would be an "evolutionary species." Something similar occurred with the two species of chimpanzee, which are separated by the Congo River: common chimpanzees live to the north, and pygmy chimpanzees or bonobos to the south. The different species of gibbon also developed through geographical separation.

However, according to Simpson's criterion the species *Homo heidelbergensis*, as we understand it, would not exist, since fossils like those of La Sima de los Huesos are the ancestors of the Neanderthals and therefore belong to the same "evolutionary species": they are simply very primitive Neanderthals. While we recognize the validity of this argument, we believe that given the morphological difference between the two types of fossils, the species *Homo heidelbergensis* should be retained for practical reasons, as is normally the case in paleontology (and particularly if we look at things from the common-sense point of view, which dictates that we should give different names to things which are distinct from one another).

On the other hand, if the Neanderthals were not human, no one would dispute that they deserve their own species. However, many researchers find it difficult to accept that human beings who buried their dead and used fire were of a different species. What is more, European Neanderthals even came to manufacture tools similar to those of modern humans (the Chatelperronian industry). Lastly, we shall see below that the industry of the first modern humans (in the Levant) was the same as that of the Neanderthals (Mousterian).

However, all of these coincidences only mean that Neanderthals were intelligent. The fact that we are the only intelligent human species currently in existence does not mean that it has always been so. We are also now the only bipedal primate species, and we have seen that in the past there were several species of bipedal hominids living at the same time. The Neanderthals represent another intelligent human species, among other reasons because the common ancestors of Neanderthals and modern humans (*Homo antecessor*) were also intelligent. There is nothing to rule out the possibility that the different intelligent human species exchanged information, produced the same type of tools, and shared the technology of fire; two intelligent species can exchange information without exchanging genes.

To put it another way, if an alien spaceship landed on our planet the day after tomorrow, the beings which came out of it would no doubt be intelligent, and we would find a way to communicate with them. Nevertheless, because they evolved in a different place, the aliens would be of a different biological species from ours. Something similar happened, albeit with fatal consequences for the Neanderthals, when our ancestors arrived on the European continent for the first time.

In any case, assuming that technical change came with biological change, as some researchers advocate, would imply four separate successive colonizations of Europe: first the manufacturers of Mode 1 (represented by the fossils of Gran Dolina) would arrive, then those who made Mode 2 (*Homo heidelbergensis*), then those of Mode 3 (the ancestors of the Neanderthals), and finally those of Mode 4 (modern humans). We believe, on the contrary, that cultural diffusion between human populations or species was more frequent than replacement of one by another, and therefore that Europe was settled only twice: first by *Homo antecessor* about 800,000 years ago, and then by our ancestors around 45,000 years ago.

There are researchers who believe they see a substantial difference, a chasm, between the Neanderthal mind and that of modern humans. Thus our superior cognitive capacity is manifested by, among other things, the fact that we are the only ones capable of devising aesthetic and symbolic concepts. We have already seen that there is debate as to whether the Neanderthals buried their dead with ceremony. For us the mere fact that they buried them already implies a ritual, and therefore a capacity for symbolism.

On the other hand, we have to acknowledge that the explosion of art occurred in the Upper Paleolithic. Before this there is some very doubtful evidence, such as the series of cuts on bones found in the Bilzingsleben deposit in Germany, over 350,000 years old, or the supposed sculpture of a woman from Berekhat Ram in Israel, which is over 230,000 years old. However, although the ancestors of the Neanderthals did not have art, nor did our ancestors at that time. The great examples of artistic expression like the paintings and engravings on walls and stone panels, and the sculptures of animals and people, are found not in the early part of the Upper Paleolithic, between 50,000 and 45,000 years ago, but only from just before 30,000 years ago. It is possible that the Neanderthals did not come to produce art simply because they became extinct before artistic activity became widespread. However, we have already noted that 34,000 years ago the Neanderthals of the Grotte de Renne had items of decoration similar to those of their modern human contemporaries, indicating that they had an aesthetic sense.

But ... if modern humans did not evolve in Europe from the Neanderthals, where did they come from? In order to answer this question we need to look at the fossil record of the Middle East.

The Levant: A Crossroads

Although Neanderthals are the result of an evolution which occurred only in Europe, at some point they left Europe and spread through Central Asia and the Levant, as demonstrated by the fossils of Teshik-Tash in Uzbekistan, Shanidar in Iraq, Dederiyeh in Syria, and Kebara, Amud, and Tabun in Israel.

But in Israel fossils of modern humans have also been found in the Skhul and Jebel Qafzeh deposits (Figure 14.2).

The Skhul rock shelter lies very close to the Tabun cave. It was excavated by Theodore McCown (1908–69) in 1931–2, and the remains of at least ten individuals with modern morphology were found there. The Jebel Qafzeh deposit produced a series of skeletons of at least twenty individuals. McCown and Sir Arthur Keith studied the Skhul and Tabun fossils during the 1930s, and concluded that they belonged to a single, very varied population of Neanderthals. However, the great paleontologist Francis Clark Howell and Bernard Vandermeersch, director of the second stage of excavations at Jebel Qafzeh (1961–80), realized that both the Skhul and the Jebel Qafzeh fossils belonged to modern humans with archaic traits. We must point out that these modern humans are not exactly like us; rather they are more like how we might imagine our very ancient ancestors to look. Some, for example, have marked brow ridges together with features exclusive to our species, like the chin on the jawbone, or a high, spherical braincase.

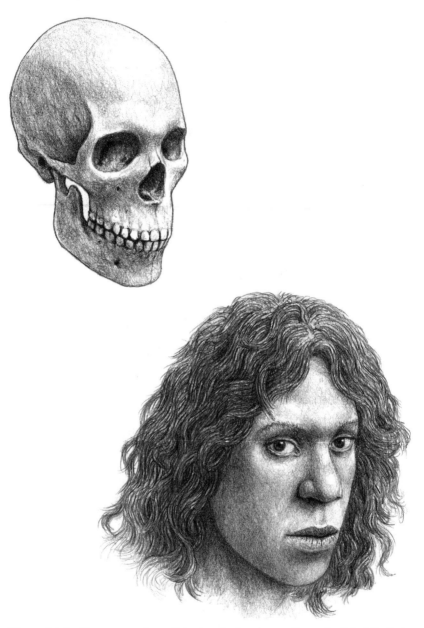

Figure 14.2   Reconstruction of the head of a woman from the Jebel Qafzeh deposit

A skull fragment (basically the frontal part of the skull) of uncertain date found in 1925 in the Zuttiyeh cave, also in Israel, corresponds to an even older population in this region. But it is not clear whether this is an ancestor of the Neanderthals or of modern humans, although the latter seems more probable.

The Neanderthals and the modern humans described here share a common class of stone technology, the Mousterian, within the great category of Mode 3 or Middle Paleolithic industries.

We might imagine that the Neanderthals were in the Levant first, and that modern humans came later and replaced them, as happened in Europe. There is even the possibility that modern humans evolved from Neanderthals in the Levant and then migrated to Europe to replace the Neanderthals there. It is more difficult to imagine that the two human types were contemporaneous in Israel, and that the Neanderthals lived in Tabun cave a few minutes away from the modern humans living in the Skhul rock shelter. This is why it is so important to know the age of both.

The human fossils from Tabun, Skhul, and Jebel Qafzeh have been dated by the TL, ESR, and uranium series methods, giving an age of about 100,000 years. However, we need to be more specific in the case of the Tabun deposit. This cave was excavated between 1929 and 1934 by the English archeologist Dorothy Garrod, who found a female skeleton (Tabun 1) and an isolated mandible (Tabun 2). The skeleton is clearly Neanderthal; the species identification of the mandible (Tabun 2), on the other hand, is more doubtful, and it might equally belong to a modern human. Both remains come, in principle, from the level dated at about 100,000 years old. However, Garrod herself admitted the possibility that the female skeleton was of more recent date and corresponded to the level above; the people of this upper level might have dug a hole to bury her and placed her in the lower level.

The Neanderthals of Kebara, Amud, and Dederiyeh are more recent, between 85,000 and 50,000 years old. These fossils are contemporaneous with the "classic" European Neanderthals. If the Neanderthal Tabun 1 skull also belonged to this period (which is not certain), and the Tabun 2 mandible was that of a modern human, then all the data would indicate that the ancestors of modern humans were the first to occupy the region, as an extension of the settling of Africa; they were then replaced by the Neanderthals. Archeologist Ofer Bar-Yosef believes that the Neanderthals colonized the Levant during a period of intense cold in Europe, which saw a displacement of the populations of central Europe toward the Mediterranean.

Finally, between 45,000 and 50,000 years ago the Mode 4 (or Upper Paleolithic) industries appear in the Middle East, and we may assume that those who made these tools were the modern humans who, having arrived from Africa, had replaced the Neanderthals, then spread through the rest of the world and replaced all other populations of other kinds of humans they found on their way.

To the Easternmost Edge of Asia

Just as our ancestors reached Finisterre, the westernmost point of Europe, by replacing the local species, modern humans also spread to the far east of Eurasia and beyond, reaching Australia more than 40,000 years ago. We do not know how much earlier, because there is some doubt as to the dating of the most ancient archeological deposits, but many authors believe that the first humans set foot in Australia more than 50,000 years ago. One interesting fact is that the Australian deposits contain Mode 3 tools, but not Mode 4, a fact which supports the hypothesis that Australia was settled before Europe. Perhaps the ancestors of the Aboriginal Australians left Africa without passing through Palestine (where the Neanderthals were at this point), instead crossing the Red Sea via the Bab-el-Mandeb Strait. They would have had to travel by sea, but they must have done this in any case to reach Australia, which was always an island (unlike Java, it had no land connection with the continent when sea level fell during the glacial stages).

Some of the Australian fossil skulls are very robust, with brow ridges and even sloping foreheads. Some authors, like Alan Thorne, believe that these traits indicate that modern Aboriginal Australians carry genes of the last Indonesian *Homo erectus*, which the modern humans would have met on their way and with whom they might have mixed. However, among the oldest fossils of Australia some, such as those from Lake Mungo (between 30,000 and 26,000 years old), are of the gracile type, while others are robust, like specimen WLH 50 (Willandra Lakes), which has not been dated but could be of similar age. Alan Thorne believes that the robust and gracile fossils have different origins (with more or less involvement of *Homo erectus*), but Peter Brown, another researcher into the origins of the Australians, believes that both types represent part of the normal variation of a single population.

In our view, the most likely conclusion is that the Australian Aborigines have no connection with the populations who lived in East Asia before the spread of modern humans. The robust Australian fossils are in fact reminiscent not of *Homo erectus* but of the first modern humans of Israel and Europe.

As in Europe, this spread of modern humans into Asia had fatal consequences for the native populations. In the easternmost area of the world inhabited by humans, the island of Java, lived the last *Homo erectus*, descendants of the populations represented by the Javanese Lower and Middle Pleistocene fossils. We have a magnificent collection of braincases of these last *Homo erectus*, found in the terraces of the Solo River in Ngandong. As far as is known, in Java as in Europe a local process of evolution, with a degree of geographic and genetic isolation, took place here. Morphologically the Ngandong braincases show clear evolutionary continuity with the older Javanese fossils, although brain size has increased (with cranial capacities ranging from 1,035 to 1,225 cc), modifying the structure of the neurocranium somewhat. The Sambungmacan braincase is generally considered a good example of a fossil intermediate between archaic specimens like those of Trinil and Sangiran and the Ngandong skulls.

It has traditionally been thought, on the basis of the fauna and geology of the region, that these human fossils from Ngandong belong to the Upper Pleistocene (less than 127,000 years old), while the Sambungmacan braincase is usually thought to be older, over 200,000 years old. The Ngandong *Homo erectus* would thus be contemporary with the Neanderthals and modern humans. If this chronology is correct, it would mean that three different human species coexisted until very recently. Carl Swisher and his colleagues have used uranium series and ESR dating to date bovid teeth associated with the human fossils from Ngandong and Sambungmacan. The result is very surprising, giving a range of ages between 54,000 and 27,000 years. But once more, we need to wait for this new dating to be confirmed, given that some eminent geochronologists, such as Christophe Falguères, disagree and believe the Ngandong and particularly the Sambungmacan fossils to be considerably older, all more than 200,000 years old.

The African Origin of *Homo sapiens*

We have already seen that the first humans who were basically like ourselves, albeit with some archaic traits, were to be found in the Middle East (Skhul and Qafzeh) between 90,000 and 120,000 years ago. In South

Africa there is a deposit of roughly similar age, known as Klasies River Mouth, which has yielded remains which are fairly fragmented but belong to modern humans. There are other fossils in Africa with modern characteristics, such as Omo-Kibish 1 and 3 (Ethiopia) and Border Cave (South Africa); these are believed to be of approximately similar age, although the dating is less certain. In North Africa fossils with modern characteristics have been found associated with a local Mode 3 tradition (the Aterian) in the Dar-es-Soltan II cave (Morocco); their age is uncertain (between 70,000 and 40,000 years old), but they are probably older than the first settling of Europe by modern humans.

Around one million years ago there is a great gap in the fossil record in Africa; this gap might well be occupied by a population of the same type as that represented by the humans of Gran Dolina (*Homo antecessor*). While during the subsequent hundreds of thousands of years the European population of *Homo antecessor* was to evolve into the Neanderthals, the African population would evolve into our species (see Figure 12.2).

If this is the case, we need to seek our furthest origins among the African fossils (of between 600,000 and 250,000 years ago), which some authors mistakenly include in the species known as *Homo heidelbergensis* (which is actually exclusively European). These fossils are those of Bodo, Eyasi, Ndutu, Salé, Elandsfontein, and Broken Hill (a group identified by the name *Homo rhodesiensis*).

Some fossils from the Asian continent are similar, particularly those from Dali and Jinniushan. Two fairly complete but deformed skulls from Yunxian (China) might also be included in this group, as might a partial skull found in Narmada (India). In theory these populations would have come from Africa and would have replaced *Homo erectus* in continental Asia, perhaps interbreeding with them. In any case, we need to know more about these fossils and their chronology in order to gain a clear understanding of what happened in Asia.

Both genetic data (which we shall discuss later) and paleoanthropological data show that our species has shown a high degree of homogeneity in its characteristics ever since it first emerged, indicating that we derive from a very reduced population which in its turn belonged to a wider, more varied species. There is a series of human fossils in Africa, less than 300,000 and more than 10,000 years old, which could already be qualified as premodern: Omo-Kibish 2 (Ethiopia), Ngaloba 18 (Tanzania), Eliye Springs and ER 3884 (Kenya), Florisbad (South Africa), and Jebel Irhoud 1 and 2 (Morocco). We ourselves derived from some African population of this epoch, although we cannot yet say which. The fossils listed here have, where they are preserved, gracile faces like our own, a large brain volume (over 1,350 cc), and braincases which have become less robust and are rounded in shape (with, for example, less angled occipital bones than in previous forms). However, as Giorgio Manzi points out, they have yet to undergo the radical transformation which, undoubtedly in just one time and place, converted a more or less low neurocranium with a flat forehead to a rounded, almost spherical neurocranium, widest in the middle and with a high cranial vault and vertical forehead – the shape we recognize as our own.

15

The Origins of Modern Humanity: The Genetic Evidence

The value and utility of any experiment are determined by the fitness of the material to the purpose for which it is used.

Gregor Mendel, *Experiments in Hybridization of Plants*

A Brilliant Idea

The origin of modern humanity has always been one of the most controversial questions in human paleontology. Many paleontologists believe, like Gunter Bräuer and Christopher Stringer, that modern humans originated in Africa between 300,000 and 100,000 years ago. From this African birthplace, our species spread through the rest of the Old World and replaced the other human species (Neanderthals and *Homo erectus*) which had appeared as a result of local evolution, in conditions of reproductive isolation, in Europe and Asia. This has been dubbed the "Out of Africa" hypothesis, in reference to Isak Dinesen's wonderful book, on which the film of the same name was based.

Moreover, the idea that Neanderthals and modern humans do not form a sequence of the ancestor-descendant type, but belong to two independent evolutionary lines which separated very early on, has also been promulgated by various anthropologists since it was first put forward in 1912 by Pierre Marcelline Boule (1861–1942).

In the preceding chapters we have followed these approaches to the origin of modern humanity and its evolutionary relationship with the Neanderthals since, in our opinion, the fossil evidence fully confirms this relationship.

However, it must also be admitted that other paleoanthropologists find, in the same fossils, evidence of a multiple and very ancient origin for modern humanity. These researchers believe that each of the different human populations which occupied the Old World after the first departure from Africa, over one million years ago, evolved in their separate geographic regions to give rise to the human populations (the different "races," to use an older term) which today inhabit the globe.

In its original formulation this hypothesis, championed by Franz Weidenreich and Carleton Coon (1904–81) among others, suggested that each human line had evolved independently and in parallel with the others. This is not a Darwinist view, since it postulates that different populations which evolve separately in different environments will ultimately converge into the same species. In order to resolve this problem the modern version of this theory, whose principal proponents are Milford Wolpoff and Alan Thorne, suggests that there was a *gene flow* between all the Pleistocene populations distributed throughout Africa, Asia, and Europe. This gene flow, it is suggested, was of sufficient magnitude to maintain the homogeneity of the human species dispersed through three continents, but not so intense as to eliminate certain specific traits which characterize humans in each region. This hypothesis is now known as the hypothesis of *multiregional origin*.

The main reason that contradictory theories of the origin of modern humans continue to exist is the nature of the fossil record. Paleontologists are trying to unravel a process which took place over hundreds of millions of years in three continents and involved thousands of individuals. In order to carry out this task they have no more than a handful of fossils, often fragmented, isolated, and dispersed in time and space. There is no doubt that the missing pieces of the jigsaw are more numerous than the ones we have.

The discovery of new fossils, the precise dating of these, and the ever-deepening knowledge of the biology of species are the tools paleontologists use to support their hypotheses. But this procedure is slow and tortuous, and depends in large measure on the pure chance of paleontological discoveries. Ideally, we need to have recourse to data from a field independent of paleontology in order to put the hypotheses based on the fossils to the test. But where can we turn to find such data?

The knowledge gained over the 20th century about the mechanisms of genetic inheritance allows us to consider a new way of approaching the problem of the evolutionary history of species. The idea is as simple as it is brilliant: it does not matter that there are few fossils of past species, since the genetic material in living species contains the key to their evolutionary history. We just need to know where to look.

We have already seen an example of this approach in the chapter on the origin of hominids and their relation to gorillas and chimpanzees. But the problem of the origin of modern humanity is different because only one of the species involved in the process survives, and we therefore only have genetic material from one of them (with one exception, which we shall look at later in this chapter). Genetic studies dedicated to clarifying the origin of modern humanity attempt to determine the genetic structure of humans today; on this basis inferences can be drawn about the how, when, and where of our origin.

The Molecules of Inheritance

The molecule responsible for biological inheritance is deoxyribonucleic acid (DNA), which carries, coded in its chemical structure, the information necessary to ensure the continuation of the species. DNA in the cells is organized into discrete units known as *chromosomes*, which in animals are housed in the nucleus of the cell. Each species has a specific number of chromosomes: humans have 23 different pairs of identical chromosomes, 46 in total.

When gametes (ova and sperm) are produced, a very special kind of cellular division known as *meiosis* occurs, as a result of which each gamete receives only one copy of each chromosome. In other words, unlike the other cells in the body, our gametes have only 23 chromosomes, one of each type. The significance of this process is clear: if the gametes carried the same number of chromosomes as the rest of the organism's cells, the egg cell which results from the fusion of two gametes would have double the number of chromosomes as the cells of its progenitors, and the number of chromosomes in a species would not remain constant over time but would double with each generation.

During meiosis a very important phenomenon known as *recombination* occurs; this consists in the exchange of fragments of DNA between the homologous chromosomes of each pair. As a result two recombinant chromosomes are obtained, whose genetic information is a mixture of that of the paternal and maternal chromosomes. The phenomenon of recombination is extremely valuable in evolutionary terms, since mixing the information received from each progenitor results in the appearance of new genetic combinations. Together with mutation, recombination is the main factor responsible for variation in organisms, the base on which natural selection operates.

The 23rd chromosome pair is different from the others because in males it is formed by two different chromosomes. It is this pair of chromosomes that holds the genetic information determining the sex of individuals. Chromosomes in pair 23 can be of two types: one an X-shaped chromosome (giving it its name, X-chromosome), and the other smaller, Y-shaped chromosome (known as the Y-chromosome). Women always have two X-chromosomes in the 23rd pair, while men have one X- and one Y-chromosome; thus women are XX and men XY.

Black Eve

Our cells obtain their energy from a series of very complex chemical reactions, most of which, particularly those which involve oxygen, take place within a series of small organelles known as *mitochondria*. In addition to performing this important role in cell life, the mitochondria are exceptional in another way: they are the only organelles in the animal cell which have their own genetic material.

The DNA of a mitochondrion is contained in a circular chromosome smaller than the chromosomes of the cell nucleus, and very similar to that of bacteria. Mitochondrial DNA is ideal for evolutionary studies for two reasons: first because all variation is due exclusively to mutation, since it is not subject to recombination, and secondly because the organelles in the egg cell come only from the maternal ovum and are transmitted through the mother's line (in the process of fertilization the sperm only contributes its nuclear chromosomes, so the egg cell consists of the ovum plus the nuclear chromosomes of the sperm).

Thus, we can trace the ancestry of a mitochondrial chromosome, from woman to woman, over the generations. The mitochondrial DNA (for which we shall use the standard abbreviation, mtDNA) in any of our cells can be identified with a single ancestor in each generation: our mother, our maternal grandmother, only one of our four great-grandmothers (the mother of our grandmother), and so on.

One might naïvely suppose that geneticists analyze the entire mtDNA of each individual to identify its peculiarities in order to compare them with those of others, but such a task would require a great deal of time, resources, and effort. Studies on variation in mtDNA are actually confined to specific regions. The regions chosen must show variation, manifested in the existence of a series of different types of mtDNA (polymorphism). The various modern human populations can be characterized in terms of the frequency in which the corresponding types occur in each of them.

Although it was not the first work published on variation in mtDNA in humans, the article by Rebecca Cann, Mark Stoneking, and Allan Wilson which appeared in *Nature* on January 1, 1987 caused a major upset in the field of studies of the origin of modern humanity. The article presented the results of an extensive study of the mtDNA of 147 individuals from 5 large groups of different types of humans (Caucasians, Asians, Africans, Australian Aborigines, and natives of Papua New Guinea). The breadth of the sample, together with the extent of the portion of mtDNA analyzed (representing about 9 percent of the total mitochondrial chromosome) contributed to the impact of the article.

The results of this study can be summed up in two main points (with which other, more limited studies, made previously by other authors, concur).

First, on the basis of similarity of mtDNA, the *existence of two major groups* was noted. In one of these only mtDNA of African origin is found, while in the other mtDNA from the other four groups appears together with some mtDNA of African origin.

The second fundamental result of the study relates to the *variation within each group*. The mtDNA of the African group showed greater diversity than that of the group which included the rest of the mtDNA. This was interpreted as evidence that the African group was the oldest of all. The use of the variation in mtDNA found within a given group as a measure of its antiquity is based on the assumption that the older the group, the more time it will have had to accumulate mutations, giving rise to more different types of mtDNA.

Finally, the authors of the article calculated the date at which the mtDNA lines separated as about 200,000 years ago, the time when the woman to whom the two lines could be traced lived in Africa.

Cann, Stoneking, and Wilson's conclusions hit the headlines in the mass media and were immediately dubbed the Black Eve hypothesis (an allusion to the African origin of our species); they were also immediately challenged.

The main objections to Cann and her colleagues' hypothesis related to the interpretation of their results and their method of estimating time.

There are two arguments criticizing the interpretation of the results. The first stresses the fact that, by the very nature of its matrilineal transmission, we might expect lines of mtDNA to be lost over time simply as a result of chance (for example, the mtDNA of women who only give birth to males will cease to be represented in the population); this alone would be enough to explain the limited variation in non-African populations.

Another criticism is based on the argument that mtDNA studies may offer a biased vision of the evolutionary history of humans, since they consider only the history of women, which might not be the same as that of the overall population.

These objections have met responses and counter-arguments, drawing in a large number of scientists. Since all the arguments put forward are reasonable, the only way to resolve the problem is to look for new evidence by studying nuclear DNA.

We will return to the issue of calculating the time elapsed since the origin of modern humans later in this chapter.

An Adam for Eve

The best way of comparing results and interpretations based on the study of mtDNA is by studying the variation of a part of the nuclear DNA which is transmitted paternally and which, like the mitochondrial chromosome, is not subject to recombination. The only nuclear chromosome fulfilling these criteria is the Y-chromosome.

The variation in some of the polymorphisms identified in the Y-chromosome can be categorized as one of a few types (or *haplotypes*) among which we can determine which is the most primitive (by comparing them with the condition found in modern apes). This is a valuable innovation with respect to earlier studies of the mitochondrial chromosome, in which the primitive type was identified on the basis of the distribution of different types of mtDNA among the populations analyzed.

The results of various analyses of different polymorphisms of the Y-chromosome all point in the same direction: modern humans had a male ancestor who lived in Africa between 100,000 and 200,000 years ago. The two most recent studies, carried out by teams led by Michael Hammer and Peter Underhill, go even further, indicating that the Khoisan people (the Bushmen) are the human population with the highest frequency of primitive haplotypes.

But in addition to these results relating to the origin of our own species, the Y-chromosome studies have given us very valuable information on other aspects of our evolutionary history. First, that there was no one single "wave" of migration from Africa, but at least two, occurring at different times. The first of these was over 50,000 years ago and resulted in the colonization of Asia and Australia; another, later "wave" arrived in Europe. These results match up with the archeological data which, as we have seen, indicate that Australia was settled by humans before Europe.

But perhaps the most striking result of comparison of the mtDNA and Y-chromosome studies is that, while the different variants of mtDNA are widespread throughout the world, the different types of Y-chromosome show more limited geographical distribution, and many of them are restricted to local groups. Luigi Cavalli-Sforza suggests a bold interpretation of these data: that it was women who spread their genes throughout the world, while the men stayed preferably in the birth group; thus, societies were *patrilocal*. It is worth remembering here Rob Foley's hypothesis (discussed in the chapter on social biology) that the first hominids, like chimpanzees, formed patrilocal societies of related males.

The Other Chromosomes

Although the analyses of the mitochondrial and Y-chromosomes lead to similar conclusions, it can be argued that these results are based on studies limited to a small part of a person's DNA and that, in addition, they are restricted to very specific chromosomes, given their particular link to one or other sex. We can legitimately question whether the "evolutionary history" of the other chromosomes would also support African origin or whether, on the other hand, it would suggest a broader view.

James Wainscoat was a pioneer of the study of the origin of modern humanity on the basis of nuclear DNA. Working on the distribution of five polymorphisms in the region of the hemoglobin gene for eight human groups, in 1986 Wainscoat and his team announced that all modern human populations derived from an ancestral African population around 100,000 years ago, which would have numbered about 600 individuals. The conclusions published by Luigi Cavalli-Sforza and his colleagues in 1988, on the basis of analysis of the distribution of 120 genetic markers (proteins coded by nuclear DNA, such as blood groups) in 42 human populations, were along the same lines. The African origin of modern humanity about 100,000 years ago was corroborated once again in 1991 by a broad study of polymorphisms in nuclear DNA carried out by two teams, led by Luigi Cavalli-Sforza and by Judith and Kenneth Kidd.

Pleistocene Park

Michael Crichton's novel *Jurassic Park*, the starting point for Steven Spielberg's film series, is based on the possibility of recovering intact DNA from fossils. When Crichton wrote his book, a series of scientific articles reporting on the discovery of dinosaur DNA through insects fossilized in amber around 100 million years ago had just been published. However, these studies have now been completely discredited and it has been recognized that the DNA found was due to modern-day contamination. Moreover, it has been demonstrated that it is impossible to find fossil DNA of such an age, for the simple reason that the DNA molecule cannot survive unaltered for this length of time; not even amber can prevent the oxidation of DNA and its consequent deterioration.

The oldest DNA which has been recovered from fossils is much more recent than that of the dinosaurs in *Jurassic Park*; it is mtDNA from mammoths preserved in the ice of Siberia between 50,000 and 100,000 years ago. The low temperatures seem to have favored the preservation of the mammoths' mtDNA, and there is no expectation that anything similar will be found outside of very specific environments such as the Siberian permafrost.

One of the main problems for the "DNA paleontologist" (apart from that of whether fossil DNA actually exists) is avoiding contamination with modern DNA through the manipulation inherent in the excavation, restoration, study, and other work carried out on fossils. The efforts made to isolate mtDNA from the famous Ice Man of the Tirol (5,000 years old) offer a good example. The first attempts were thwarted by the presence of contaminating modern DNA, and the techniques had to be refined in order to find mtDNA whose authenticity was not in doubt.

Although the results obtained from the study of the mtDNA from the Ice Man were modest (his mtDNA was shown to belong to a type characteristic of central European populations), the study did offer some very valuable lessons on the techniques and controls necessary in order to get around the problem of contamination with modern human DNA.

Using these new procedures, an international team (comprising Matthias Krings, Anne Stone, Ralph Schmitz, Heike Krainitzki, Mark Stoneking, and Svante Pääbo) embarked on the search for mtDNA in Neanderthal fossils. They took samples from the type specimen of this species, the *Neanderthaler* skeleton. The study was very carefully planned and surrounded with all possible precautions, both to avoid contamination as far as possible and to prevent pointless destruction of a fragment of this valuable fossil. The 3.5-g sample was taken from a place theoretically inaccessible to contamination, the inner part of one of the bones – specifically, the right humerus.

The next step in the study was to analyze the structure of the fragments of what was presumed to be Neanderthal mtDNA. The DNA molecule is made up of thousands of smaller units known as nucleotides, of which there are only four different types in DNA. The structure of a fragment of DNA is simply the sequence of nucleotides which make it up, so this part of the work consisted of determining the nucleotide sequence in the different fragments of DNA found in the sample taken from the fossil.[1]

Many of these fragments showed sequences which were identical over some part of their length: in other words, they overlapped. This made it possible to reconstruct the sequence of the original segment from which they came. Thus, after three months of intensive work, the sequence of a segment of 379 nucleotides was reconstructed, on the basis of 123 different fragments. This sequence corresponds to region I of the control segment of mtDNA.

The researchers compared the mtDNA sequence from the fossil with 16 types of mtDNA from chimpanzees and 1,986 types from different modern human populations, and obtained some extremely valuable results. First, they determined that on average humans and chimpanzees differ in 55 bases of the base sequence, while the mtDNA analyzed showed an average difference of 27 bases from modern humans. Moreover, they established that within the modern human sample the average difference was only 8 bases. Thus, the fossil mtDNA is sufficiently similar to our own to recognize that it comes from a human being, but different enough to reject the suggestion that it belongs to a modern human, thus ruling out the possibility of contamination and confirming it as an authentic fragment of Neanderthal mtDNA.

The fact that the average difference between Neanderthal mtDNA and that of modern humans is more than three times greater than the average difference among modern humans (27 bases compared to 8) led the

---

[1] A series of fragments of mtDNA, representing two types of sequence, was obtained from the Neanderthal fossil. One of these sequence types was attributed to modern mtDNA, present through contamination, while it was thought that the other might correspond to the authentic mtDNA of the fossil. It is striking that despite all the precautions taken, which theoretically rendered the presence of contaminating DNA impossible, it did appear in the analyses. This fact indicates how difficult it is to avoid contamination in this type of work, and throws into question the results of other, less rigorous studies.

researchers to conclude that the two lineages separated a very long time ago.[2] In order to establish when this separation occurred, the researchers used a mutation rate determined on the basis of the average number of changes observed between modern humans and chimpanzees (55) and the time which is assumed to have passed since the two lines separated. Their calculations gave an age of between 550,000 and 690,000 years for the Neanderthal/modern human separation, and between 120,000 and 150,000 years for the origin of present-day human diversity.

The conclusions about the age of modern humans concur with those derived from studies of mtDNA, the Y-chromosome, and other nuclear chromosomes. One secondary result of the analysis is that the mtDNA of modern humans of African origin again appears ancestral to that of the other modern human populations.

In addition to reinforcing the theory of a single, African origin for modern humanity, this study has another noteworthy feature of particular relevance. The separation between Neanderthals and modern humans is traced much further back than the majority of authors working on fossil studies had suggested up to this point. However, this separation date is compatible with the results of our investigations of the human fossils of Atapuerca, both from La Sima de los Huesos and from Gran Dolina level 6. As we have already noted, the Gran Dolina fossils, which are close to 800,000 years old, represent the ancestor species of both the Neanderthal and the modern human lines, *Homo antecessor*.

Perhaps the discovery of Neanderthal fossil mtDNA will give rise to a new novel, a tale of cloning Neanderthal individuals. However, one of the conclusions which can be drawn from our better knowledge of fossil DNA is that we will never find sufficient nuclear DNA to allow us to dream of the Pleistocene version of Crichton's book.

Fossils and Molecules

It has frequently been suggested in the popular media that genetic studies just by themselves have revealed the "how," "when," and "where" of our

---

[2]   Nevertheless, some comparisons of mtDNA within modern humans gave a greater average difference (up to 24 bases) than that obtained from comparison of Neanderthal mtDNA with that of certain modern humans (20 bases). Thus the results of the study are not as convincing as has been suggested, but should rather be considered statistically reliable.

origin. However, as we pointed out at the beginning of this chapter, the main hypotheses as to "when" and "where" have been based on the study of fossils, and were put forward before the genetic studies began.

In this context, the genetic studies have reinforced the hypothesis of the African origin of modern humans (the "Out of Africa" hypothesis). Moreover, genetic analyses have also corroborated the opinion of many paleontologists that Neanderthals were not either direct or indirect evolutionary ancestors of modern humans, but that the two human species share, far back in time, a common past.

One important area which genetic studies have been essential in illuminating is the "how" of our origin. The narrow genetic range noted in non-African human populations indicates the existence of an evolutionary phenomenon known as a "bottleneck" at the point when our species left Africa. "Bottlenecks" are created when a biological population originates from a relatively small number of individuals who carry only a fraction of the genetic diversity of the mother population. As a result, the population derived from this small group receives only a part of this diversity.

We can see a good example of this phenomenon in the Afrikaner population of South Africa, which largely originates from a handful of Dutch pioneers who arrived there in the 17th century. There is evidence that one of those colonists, who arrived in 1688, suffered from a rare genetic disease known as *porphyria*. And we find that today, the frequency of this disease in the South African Afrikaner population is several hundred times higher than in any other human population.

Similarly, the high degree of homogeneity of the non-African populations tells us that the modern human groups which colonized Asia and Europe were formed by a small fraction of the original African population.

The number of pioneers can also be fairly reliably established. A team of geneticists including Spanish-born researcher Francisco Ayala have analyzed the present-day variation, in modern humans, in the genes responsible for the main histocompatibility complex, or HLA system, which is involved in defense against microbial invasion through its capacity for recognizing proteins alien to the organism.

These genes are located on chromosome 6 and are extremely varied, allowing the HLA to recognize a large number of molecules as alien. The range of variation observed in modern humans indicates that the "bottleneck" through which our ancestors passed was not too tight, since otherwise there would be much less variation in the HLA system. According to these researchers, the number of settlers who left Africa must have been greater than 500, and was probably around 10,000.

While genetic studies have been useful in comparative analysis of pale-ontological hypotheses on the "where" of the origin of modern humans, and have been particularly valuable as a source of information on the "how," they have not been so eloquent on the question of *when* we appeared.

The measurements of time calculated in the genetic studies, known as "molecular clocks," are always problematic because they are based on a series of highly debatable assumptions, which we have already discussed in the chapter on the origin of hominids (the assumption that the rate of mutation is constant, or that the regions of DNA studied are neutral with regard to the action of natural selection).

Perhaps the most serious criticisms of "molecular clocks" are those which cast doubt on the rates of mutation used to calculate the moment when modern humans emerged. In their original article Cann, Stoneking, and Wilson recognized that time elapsed cannot be reliably established on the basis of variation in mtDNA alone. Like all other authors, they adjust their "molecular clocks" by using the fossil record, and therefore their results cannot be invoked to compare hypotheses based on fossils. In this respect paleontology and genetics are not independent sources.

At the beginning of the 21st century, our view of the origin of modern humanity is much more complete than might have been expected even 20 years ago. We are much more certain of the place and time when our species emerged, and we have made a great deal of progress in clarifying how the process took place. Contrary to some expectations, molecules have not replaced fossils in the study of human evolution. The different approaches of paleontology and genetics have enabled us to consider the problems from different perspectives, enriching our knowledge of the evolutionary history of the species *Homo sapiens*.

Standards of Beauty

Up to now we have been considering our species' past, but what of its future? Many believe that, now we have machines enabling us to depend less and less on our physical strength to survive, and more and more on our intelligence, the corresponding organs will be affected in the future. Thus the human being of tomorrow is frequently represented as having an atrophied body and a very big head, or rather, with an overdeveloped brain, since the face and teeth are also depicted as reduced – in short, a very unattractive individual, in terms of the ancient Greek canons of beauty. These theories usually give no precise date as to when we will become such unathletic, although highly intelligent creatures, but it seems that at the rate we are going the standards of beauty will have to change very quickly (goodbye to 36–24–36!).

But is it possible that natural selection is still acting on us, and can determine the future course of human evolution? There is absolutely no doubt that our species is subject to natural selection, like all others. Individuals with serious genetic defects do not live to adulthood and do not reproduce, and many die in utero and are not even born. But this normalizing selection, which eliminates extreme individuals, does not modify the species. For the species to evolve in a given direction takes a very long time, and it also requires individuals with specific characteristics to reproduce more than others, something which appears not to be happening, or not on any great scale. Moreover, technology allows us to adapt rapidly to life in all kinds of environments, including the moon, without changing our morphology. Adaptation by natural selection is much slower (and more limited). To take just one example, thanks first to writing and now to information technology, our brain does not need to grow any more in order to accumulate and process more information.

On the other hand, our species is already very numerous, and there is therefore a great genetic inertia or resistance to changes, which are diluted like drops of water in the ocean. An interesting phenomenon is beginning to arise, which will go against the trajectory of humanity over the last few thousand years: human populations, which became isolated from one another and evolved into different races, are beginning to mix and exchange genes, and it is thus certain that there will be new genetic combinations. But this does not mean that the species will change substantially in the near future.

One final, disturbing factor should be mentioned. Since we discovered artificial selection 10,000 years ago with agriculture and animal husbandry, we have always been capable of modifying ourselves just as we have modified breeds of animals. It does not appear that this has happened on any major scale. Now, however, with our knowledge of genetics, we are beginning to have the real possibility of modifying our own genes much more rapidly and radically than with artificial selection (and even more rapidly and radically than natural selection). Genetic manipulation, which can free us from defects and disease, could also be used for other purposes. But in itself it is simply a tool, like all those that science makes available to us, and it is our responsibility to control it.

16

The Origins of Human Language

Thus it was necessary for the said hole in the Adam's apple to close when meat is swallowed that it does not enter the throat, which would be a vexing and harmful thing, of which we have experience every day when we hurry in swallowing and something enters our throat, because there follows from this a most troublesome and grievous coughing.
Bernardino Montaña de Monserrate, *Libro de la anathomia del hombre* [Book of the Anatomy of Man]

King Solomon's Ring

Human beings are the only organisms which speak. That is to say, we transmit to our fellows, and receive from them, all kinds of new information, deliberately encoding our messages in combinations (words) of pre-established sounds (syllables). Other animals are only capable of exchanging very specific information on certain aspects of their life, using a limited system of sounds and gestures which are not intentionally encoded.

Konrad Lorenz, in a delightful book entitled *King Solomon's Ring*, makes reference to the legend that King Solomon possessed a ring which enabled him to talk with beasts. Lorenz boasted that he too was able to understand the simple vocabulary of animals, without needing a ring; but he added that animals do not have a true language; rather, each individual innately possesses a code of signals formed by expressive sounds and movements of which another individual of the same species has a similarly innate understanding. However – and this is where they differ fundamentally from humans – animals emit these signals automatically when they find themselves in a particular state of being, even if there is no one to witness them. Lorenz expressed this idea by saying that with their sounds animals are emitting not "words" but "interjections." The *Merriam-Webster Dictionary* defines an interjection as "a cry or inarticulate utterance expressing an emotion." And this is exactly what animals are emitting, according to Lorenz.

These ideas appear to be true for the majority of animals, but they are perhaps not quite so accurate in the case of primates. Dorothy Cheney and Robert Seyfarth have studied vervet monkeys (*Cercopithecus aethiops*) in the wild in Africa, observing that in addition to emitting signals, in the form of sounds and gestures, which communicate their motivation or state of mind, these animals also inform one another about certain aspects of the environment. For example, they have different sounds to warn of the presence of different predators – snake, eagle, or leopard. The reactions that these calls stimulate in the hearers are different in each case: if the warning is of a leopard they climb trees; if the predator is an eagle they hide in the bush or look up; and if a snake is signaled they stand up and scan the grass. Thus, each call has a distinct meaning which triggers a different response; these are not simply cries of fear in reaction to the presence of a predator.

Moreover, Cheney and Seyfarth's research on other calls related to the social life of the vervet monkey has revealed that these monkeys associate sounds which have similar meanings, even if they are acoustically very different (just as we would with the words "car" and "vehicle").

In the 1960s and 1970s the idea of direct communication with the animals most like us, chimpanzees and gorillas, was taken very seriously in some research programs. Since chimpanzees and gorillas are physically unable to pronounce words, the task of communicating with us was facilitated by teaching them sign language, which they were able to reproduce. The chimpanzees and gorillas proved to be outstanding pupils, and revealed their capacity for associating ideas which we express in words with gestures, or with tokens of different shapes, with drawings and colors (icons).

The bonobo Kanzi (whose efforts at carving stone tools we described in a previous chapter) understands more than 150 words of spoken English. And Kanzi was not the only chimpanzee to show a degree of linguistic capacity. Washoe was the first chimpanzee to learn a series of sign-language gestures (132 signs during a little over 4 years of training); Sarah showed indications of an ability to detect the order of objects used as words; and Lana appropriately completed sentences made up of characters representing words, paying attention to the meaning and order of the words.

One classic question relating to animals' systems of communication is whether they can deceive one another for their own benefit. Knowing how to lie would make them more "human," as it would indicate that they are not simply automatons, but are able to control what they express. In fact, chimpanzees in the wild have been observed to deceive their companions on many occasions, in a wide variety of contexts, with their gestures, posture, and facial expression. Our sins are also theirs.

The results of all these investigations are extremely valuable, because they have revealed an incipient linguistic skill in primates which was previously denied. However, they are disappointing in that none of these primates has communicated any relevant information on themselves. Vervet monkeys appear to have a limited repertoire of "words" which they use in very specific situations, and chimpanzees have demonstrated that they are very skillful in manipulating symbols, and can even be consummate liars, but that is all. Perhaps the myth of King Solomon's ring is, after all, just that – a myth.

Given that human language is so different from that of our living relatives, the question of its origin and development can only be tackled from the viewpoint of paleontology.

Language and Brain

Paleoneurology is a science which attempts to determine the mental capacities of a fossil species from the impressions left by the brain on the inner surface of the skull. There are two areas of the cerebral cortex, both in the left hemisphere of the brain, that are closely related with speech in humans (see Figure 8.1). *Broca's area*, located in the third frontal gyrus (at the level of the temple), is responsible for syntactic construction and planning: in other words, it translates messages into an ordered sequence of movements of the muscles involved in the production of speech. A lesion in this area disrupts the capacity to speak and write, but not the individual's understanding of spoken language, and the affected person will still be able to read. *Wernicke's area*, on the other hand, located between the upper temporal gyrus and the parietal lobe (a little behind and above the ear), is responsible for encoding and decoding messages. A lesion in Wernicke's area renders the individual incapable of correctly understanding and producing language, whether spoken or written.

According to Philip Tobias, the lower region of the parietal lobe, the part related to Wernicke's area, is more developed in the *Homo habilis* fossils from Olduvai than in australopithecines, *Paranthropus*, and apes. Moreover, Broca's area is markedly enlarged both in *Homo habilis/Homo rudolfensis* and in *Homo ergaster*. This area is much more developed in the first humans than in australopithecines and *Paranthropus*, in which it is only sketchily present.

Thus, the regions of the cerebral cortex most directly related to the production of human language were already well developed in the first representatives of our genus. Does this mean that these humans already possessed the capacity of speech? Although this is the conclusion of the majority of specialists who have studied brain casts from primitive hominids, there are those who oppose this view.

In his book *The Wisdom of the Bones*, Alan Walker, who led the study of the Turkana Boy, rejects the possibility that this individual, and by extension all those of his species, was able to speak. Walker bases his conclusion on the fact that within the thoracic vertebrae of Turkana Boy the vertebral canal is very narrow. This configuration is common among apes, but not in modern humans, who have a much wider vertebral canal. In view of the reduced diameter of the vertebral canal, Walker argues that the spinal cord of the fossil specimen contained fewer neurons than that of modern humans, so Turkana Boy's thoracic region would have been less innervated than our own. The only plausible explanation of this fact, according to Walker, is that the thoracic musculature involved in breathing was not capable of performing the precise inhalations and exhalations that human speech requires. So how do we explain the high level of development of Broca's area reflected in the inner surface of the braincase of Turkana Boy? On the basis of results obtained from modern techniques used to explore brain activity (specifically *positron emission tomography*, or PET), which show that the area of the cerebral cortex surrounding Broca's area is also related to manipulation of the right hand, Walker suggests that the development of this area in the first humans was an adaptation related not to speech, but to dressing stone.

To sum up, although different studies of the cerebral cortex of the first hominids concur that the areas connected with language (particularly Broca's area) are more fully developed in the first humans than in australopithecines, *Paranthropus*, and apes, they do not agree on the physiological significance of this development. The solution to this problem could come from research into the anatomy of the phonetic apparatus of fossil hominids.

The Choking Primate

The second route of investigation into the origins of language is to study the set of organs responsible for the emission of the sounds that make up human speech. Although this might at first appear to be of secondary importance, given that the mental capacities required to possess a complex language do not depend on the physiological capacity for producing it, it is nevertheless true that one cannot compose music for instruments that do not exist.

The sounds on which human language is based are produced and modulated in a series of cavities which make up the upper stretch of the respiratory tract and are known collectively as the *vocal tract*: the larynx, pharynx, and the nasal and oral cavities (Figure 16.1).

In all mammals apart from adult humans, the larynx is situated high in the neck, almost at the exit of the oral cavity. This high position allows the larynx to connect with the nasal cavity during the ingestion of liquids, which thus pass from the oral cavity to the alimentary canal without any interruption of breathing. In other words, any mammal is able to breathe through its nose while drinking. However, in adult humans the larynx is positioned unusually low in the neck, so that although we are mammals, we are unable to breathe while drinking.

The importance to a mammal of being able to breathe through the nose while drinking is clear if we think of young creatures who are being breastfed. For this system of feeding to be efficient, an infant must be able to breathe as it suckles.

It will not have escaped the perceptive reader that our babies are also able to breathe through the nose while suckling or drinking from a bottle. This shrewd reader will thus have correctly surmised that in nursing humans the larynx is in the same position as in other mammals (Figure 16.2). In our species the larynx descends at the age of about two years. From this point, not only do we lose the capacity to breathe while drinking, but the unusual position of the human larynx makes it possible for food to obstruct the respiratory tract, since the epiglottis is not able to cover it completely (Figure 16.1). Choking is no joke – we can die from it.

But if our upper respiratory tract has lost efficacy in this respect (and also for breathing and sense of smell), what is the compensation? The answer is that our species has a pharynx larger than that of any other mammal, giving us the capacity to modulate a wide range of different sounds.

The Production of Speech

Contrary to popular belief, the majority of the basic sounds which make up human speech do not originate directly in the vocal cords. In the production of vowel sounds (and also of voiced consonants, but for the sake of simplicity we will refer only to vowels in what follows), the vocal cords open and close rapidly with periodic breaths of air to produce a "base" sound, or *laryngeal tone*. This is always the same, regardless of the vowel we are pronouncing.

The laryngeal tone is made up of one main frequency and a series of "accompanying" frequencies or harmonics. If the set of cavities above the larynx (the pharynx, nasal cavity, and oral cavity) were not involved

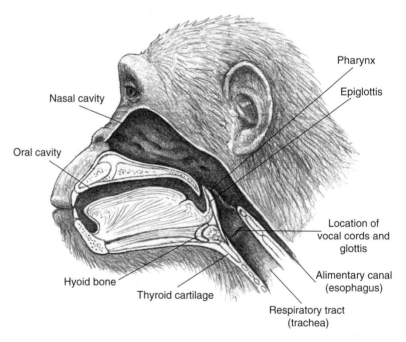

Figure 16.1 (cont'd)

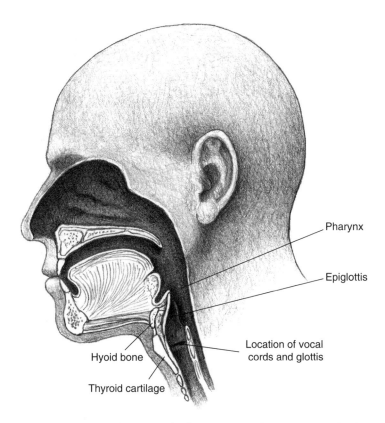

Figure 16.1 Section of the head of a common chimpanzee and a human, showing the upper airways. On its inward passage towards the lungs, air usually enters via the nasal cavity. From here it passes to the pharynx, a channel common to both the alimentary canal and the respiratory tract, serving as a passage for both air and food and fluids. The pharynx continues downward to the point where the alimentary canal and respiratory tract separate. The entrance to the respiratory tract is formed by a cartilaginous vessel known as the larynx. The upper part of the larynx is formed by the thyroid cartilage, which forms an easily distinguished protuberance in the neck (particularly in men), the "Adam's apple." In the lower part of the larynx there are two bands of muscle covered with an elastic sheath, the vocal cords. The space between the vocal cords is called the glottis. The primary function of the vocal cords is to block the glottis when they close together, thus preventing foreign bodies entering the respiratory tract. The epiglottis, a spoon-shaped cartilage located above the larynx, also contributes to this function. When we swallow or drink, the larynx rises so as to position the glottis beneath the epiglottis, partially blocking the respiratory tract against the passage of food and fluids. As a result, breathing is interrupted for the duration of the time the larynx remains closed.

in the production of human speech sounds, the sound we heard would consist mainly of the note corresponding to the principal frequency of the laryngeal tone; most of its harmonics are too weak for us to hear them.

However, this is not what actually happens, owing to the phenomenon known as *resonance*, whereby a body (resonator) is able to produce vibrations as a result of the vibration of another adjacent body. This is why window glass vibrates (resonates) in response to traffic noise. However, not all bodies which act as resonators reproduce the same vibrations: each type of resonator responds to certain particular vibrations. Thus if we pass a sound made up of various frequencies (as the laryngeal tone is) through a resonator, the resonator will reproduce, and even amplify, some frequencies but not others. The result will be an acoustic filter, where some frequencies have been amplified and others silenced; the resulting sound will be different from the original. Depending on the resonator we use, we will get different sounds from any given base sound.

The human vocal tract can adopt different configurations, each of which acts as a different resonator which filters the laryngeal tone produced by the vocal cords in a specific way, giving rise to the different vowel sounds. In order to generate this filter, narrowing and widening has to take place at specific distances from the source of the laryngeal tone, both in the pharynx and in the oral cavity.

For example, in order to produce the sound [a], the tongue flattens in the oral cavity and shifts backward to narrow the pharynx. At the same time, the larynx rises to shorten the distance between the vocal cords and the narrowed zone (pharynx) and the widened zone (oral cavity). The opposite occurs when the sound [i] is produced. In this case the tongue rises, narrowing the oral cavity, and the pharynx widens. In the case of [u], both the oral cavity and the pharynx widen, while the posterior part of the tongue rises to narrow the posterior part of the palate.

For this complex articulatory coordination to take place, the oral cavity and the pharynx need to be able to operate as two independent tracts, and the pharynx therefore has to be of a certain length and also has to be positioned at an angle to the oral cavity. In other words, the larynx needs to be located low in the neck (Figure 16.1).

Vowel sounds formed in this way are modified in the oral cavity through movements of the tongue, lips, and soft palate (where the uvula is located), producing consonants. For the tongue to reach the correct positions for production of consonants rapidly and precisely, it needs not to be too long. This is because some of the muscles which move the tongue insert into the hyoid bone, located in the lower and posterior part of the tongue (Figure 16.1). The longer the tongue, the longer these muscles are, and the slower and less precise their movements.

It is easy to estimate the length of the tongue from a fossil skull, since it is proportional to the length of the hard palate. However, deducing the position of the larynx is a different matter. As we have already noted, the larynx is formed of cartilage and supported by muscles and ligaments, none of which fossilize.

The Fossils Speak

Since the mid-1970s, linguist Philip Lieberman and anatomist Jeffrey Laitman have led a series of studies aimed at reconstructing the morphology of the upper respiratory tract of fossil hominids. As a result of their research they have concluded that a series of features in the base of the skull can be used to determine the position of the larynx in the neck and thus to establish the phonetic capacities of fossil hominids. The characteristic most widely accepted by the scientific community as an indicator of these capacities is the degree of flexion of the base of the skull.

If we were to cut through the human skull at its median line, or plane of symmetry (the plane which divides the skull into two equivalent halves), we would find that its lower edge shows a marked upward flexion between the *foramen magnum* and the posterior part of the palate. However, newborn humans and apes in general show very little skull base flexion. In humans this flexion increases during childhood, reaching its maximum in adulthood.

Since human newborns and apes both have a flat skull base, combined with a high position of the larynx, and given that in humans the descent of the larynx is accompanied by an increase in skull base flexion, there appears to be a clear relationship between the position of the larynx and the degree of flexion. This relationship has also been observed in experiments on rats, in which an increase in skull base flexion was generated surgically.

On the basis of this premise, Laitman and his colleagues have carried out various studies on different fossil hominids, and have drawn a number of conclusions about their phonetic apparatus. According to these researchers, in australopithecines, *Paranthropus*, and *Homo habilis* the larynx must have been located high in the throat and their phonetic capacities would have been similar to those of chimpanzees. However, they found that the skulls from Broken Hill and Steinheim (from the

Middle Pleistocene of Africa and Europe, respectively) showed flexion of the skull base, implying a low position of the larynx and phonetic capacities similar to our own. With regard to the Neanderthals, they came to the conclusion that their phonetic apparatus allowed them to articulate only a limited repertoire of vowel sounds (which did not include [a], [i], or [u]), meaning that their spoken language would have been rudimentary and slow.

However, we have studied the base of the skull in the only specimens of *Homo habilis* (OH 24) and *Homo ergaster* (ER 3733) in which this part of the skull is well preserved, and these show flexion values greater than those of australopithecines, chimpanzees, and gorillas. These results suggest that the phonetic apparatus of *Homo habilis* and *Homo ergaster* was already similar to our own (albeit that in *Homo habilis* the palate was proportionately as long as that of chimpanzees, indicating a more limited repertoire of consonants). Thus they reinforce the hypothesis that all species of our genus have had speech, given that, if these first humans did not talk, the low position of their larynx is difficult to explain in terms of natural selection.

Moreover, many researchers find it difficult to accept that the Neanderthals could have had a reduced capacity for speech compared to that of their ancestors (for example, the Steinheim fossil).

In response to these criticisms, Laitman has suggested that the phonetic capacities of Neanderthals were reduced as a result of an adaptation more important for their survival, that of modifying their upper airways to warm and humidify the cold, dry air of the ice ages: breathing is more important than talking. Moreover, in their book *In Search of the Neanderthals*, Chris Stringer and Clive Gamble postulate that the ancestors of the Neanderthals (Steinheim and Petralona) did not have a spoken language like ours, despite having the anatomical structure for it, because of the mental limitations of their relatively small brains.

However, the theory that Neanderthals were not able to speak as we do began to look doubtful when, in the mid-1980s, paleoanthropologist Jean-Louis Haim announced that the skull of the Neanderthal specimen known as the "Old Man," from La Chapelle-aux-Saints, had been poorly reconstructed by the first researchers and that his new reconstruction showed a greater degree of skull base flexion. This was confirmed by David Frayer, who measured the flexion in the new reconstruction of the specimen from La Chapelle-aux-Saints and found that it was similar to that of a series of medieval skulls. Given that this fossil was one of those studied by Laitman, his results were thrown into question.

Moreover, in 1989 a hyoid bone belonging to a Neanderthal specim the only one so far published for a fossil hominid, was found in the Isr deposit of Kebara. As we have already noted, the hyoid bone inserts i the musculature of the tongue and its position in the throat is closely related to that of the larynx. The Kebara hyoid bone shows a morphology and dimensions comparable with those of the hyoid bone of any modern human, leading the team of scientists who studied it, led by Baruch Arensberg, to conclude that Neanderthals were anatomically just as capable of speech as modern humans. This conclusion has been challenged by Lieberman and Laitman, who assert that the morphology of the hyoid bone is not relevant in determining the phonetic capacities of hominids. Unfortunately, no skull has been found at Kebara, and it is thus not possible to compare the morphology of the hyoid bone with the degree of skull base flexion.

Research into reconstructing the phonetic apparatus of the Neanderthals is thus currently at an impasse. Some researchers hold that the studies of skull base flexion are not valid because they were carried out on poorly reconstructed specimens, and they prefer to trust the results of analysis of the Kebara hyoid bone. Others deny the validity of these studies and continue to uphold the conclusions drawn from the analysis of skull base flexion in the Neanderthals.

The only way out of this situation is to find new fossil material which includes both intact skull bases and hyoid bones. It might seem that a find of this nature is almost impossible, since it involves the highly unlikely combination of finding both an intact skull and a hyoid bone (only one is known in the entire hominid fossil record, that of Kebara). However, such a discovery has recently been made in La Sima de los Huesos deposit in Atapuerca, where we found a skull with a virtually complete base, Skull 5 (see Figure 12.3), as well as the majority of two hyoid bones. We will have to await the completion of research currently under way on this extraordinary fossil material in order to learn more about the origin of human speech.

Group Selection and the Extinction of the Neanderthals

In today's world we are used to the idea that communication and information are the key to progress, and the basis of our current technological development. Thus it appears to us that possessing an articulated language gives us an indisputable superiority over other living beings in the struggle

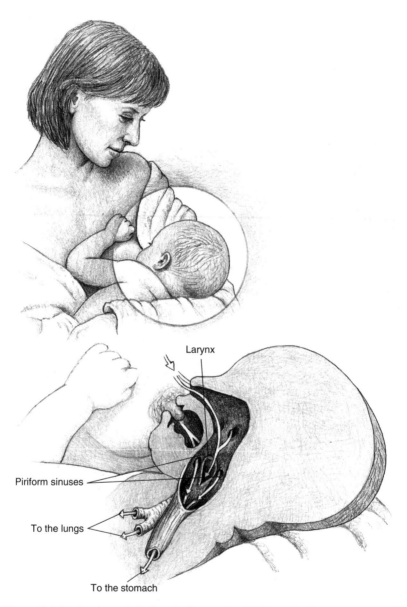

Figure 16.2 Section of the head of a nursing infant. The larynx is connected to the nasal cavity and liquid passes into the alimentary canal through the piriform sinuses. It is thus possible to drink and breathe at the same time. In adult humans, the low position the larynx means that it has no a direct connection to the nasal cavity, and breathing therefore has to be interrupted during the ingestion of liquids.

for existence. But on reflection, it is easy to see that linguistic ability has no advantage for a lone human confronting nature with only his own resources: this characteristic only has meaning within a group to which the individual necessarily belongs. Language is a property not of the individual, but of the collective. It is not the individual who communicates well; it would be more appropriate to say that the group has good communication. The capacity to share and transmit information between individuals and between generations by means of language confers a great adaptive advantage on the group as a whole, but not on the isolated individual.

This fact, which appears obvious to us, poses a serious problem for evolutionary biology, because the theory of evolution through natural selection – the Darwinian theory that is universally accepted, with minor variations, by all scientists – is based on competition among individuals. We have already noted that some authors have proposed a reductionist version of natural selection, occurring at a level below that of the individual, such as the level of the genes. We have seen in this book that competition can sometimes occur at the level of sperm – again, below the level of the individual. Darwin himself put forward the mechanism of natural selection in order to explain certain characteristics of individuals which cannot be understood from the point of view of simple competition for environmental resources and the struggle for survival; after all, what matters is surviving in order to reproduce.

However, the appearance of properties which characterize groups, such as communication, poses the problem of natural selection at the *supraindividual* level, the level of the group. Thus, language was selected because groups with a higher level of internal communication were more competitive, more efficient in exploiting the resources of their environment, and displaced other groups. In other words, an individual with the capacity to produce and understand articulated language may be no more competitive than another individual who does not have this ability, but a group with language is more competitive than a group without.

It has been known since ancient times that competition between groups of the same species occurs among social mammals, including primates. Chimpanzees, to take a case in point, defend their territory and sometimes invade that of other groups, or are invaded by neighboring groups. Not only does communication serve to increase the cohesion and efficiency of groups ("information is power," as we say in our society), but it appears that behaviors of social cooperation ("social altruism") within the group are extremely important in selection between groups. Social cooperation extends to many areas, such as defense of territory or resources, defense against predators, group hunting, caring for other adults' offspring, sharing food, and so on.

However, while competition between groups is easy to understand, formulating a realistic model of group selection is a more complex matter. If we start from the premise that cooperative behavior and the mental and phonetic capacity for language have a genetic basis, we can approach the problem by considering the existence of a gene *for cooperative behavior* and a gene *for language* (this is, of course, an oversimplification, which we hope the discriminating reader will forgive).

In order for selection between groups to occur, there must be a high level of genetic homogeneity within each group – in other words, a high level of consanguinity. To put it in more technical terms, the variability within each group must be much lower than the variability between the different groups. Groups in which the language and cooperation genes occur frequently will be more competitive than others; since cooperation and communication are not governed by the law of all or nothing, it would in fact be the greater or lesser capacity for language and cooperation that made the difference. In the case of social cooperation, there is the additional problem that "selfish" individuals profit from the efforts of others without expending their own energies, and could thus benefit from natural selection and leave more genes for the succeeding generation, with the result that "altruistic" behavior would never become dominant. One solution is that there might be mechanisms of social rejection which prevent "selfish" individuals from reproducing, but then we are faced with the question of how these mechanisms were selected, and we are back to square one.

Chimpanzees form social groups in which the males are related but not the females, who move from one group to another. As we have seen, we can picture something similar happening among the first hominids. Thus, the social group is not a closed reproductive group, and hence genetic uniformity is lost when females (genes, in short) are imported from other groups. Even so, we can formulate mathematical models which make selection at group level viable under certain conditions. But it is also possible to imagine that our ancestors came to form larger social units, which functioned as reproductive units and also competed with one another. Whether or not this is the case, at present we have no better mechanism than group selection to explain one of our most important characteristics: articulated language.

Another question for which selection at group level appears the only practical explanation is the replacement of the Neanderthals by modern humans. It does not seem reasonable to suggest that this substitution took place through selection at individual level. As we have already noted, the

Neanderthals were physically stronger and their anatomy was better adapted to the European climate (see Figure 13.1), and we therefore have to believe that individually they were as well suited as modern humans to their environment (if not more so). It seems virtually certain that the success of our ancestors was based on some group characteristic, but which one?

Many authors believe that the absence of a true language among Neanderthals was one of the reasons, if not the main reason, for their replacement by modern humans. To put it simply, the rudimentary language of the Neanderthals limited their social complexity, restricting their capacity to transmit information essential for exploiting the resources of their environment. And when modern humans appeared on the horizon, well equipped with their sophisticated language, the Neanderthals were doomed to extinction.

Although the hypothesis of the linguistic superiority of modern humans is an attractive one, we have already seen that it does not fit with the fossils. Apart from the fact that it is not clear what kind of language Neanderthals had, we have already noted that they were not "swept away" by modern humans in a rapid, universal process: the replacement of one by the other took place over about 10,000 years. If modern humans had such an overwhelming advantage in terms of social complexity and exploitation of the environment, why did it take them so long to replace the Neanderthals?

We believe that what enabled our ancestors to displace the Neanderthals was not the possession of a qualitative advantage, the type of language, but rather the higher level of development of their capacities to exploit resources: to put it simply, they had more of the same.

The Crooked Lines of Natural Selection

The men who discovered natural selection had an exemplary relationship. There was never room for professional jealousy between Charles Darwin and Alfred Russel Wallace; on the contrary, their relationship was one of mutual respect and warm friendship. When, on April 26, 1882, British society sought to recognize the importance of Darwin's work by burying him in Westminster Abbey, alongside the tomb of Sir Isaac Newton, Wallace was among the three close friends who, together with members of the aristocracy and the government, carried Darwin's coffin.

Nevertheless, and despite their good friendship and mutual admiration, Darwin and Wallace held opposing views on the origin of some of the most notable characteristics of our species, such as intelligence and speech. While Darwin saw these traits as one further result of an evolutionary process governed by natural selection, Wallace attributed them to supernatural causes, placing the origin of human beings beyond the action of natural selection.

One of the most profound arguments that Darwin advanced to support the theory of evolution was the existence of "botched jobs" among living beings. If organisms were the result of a direct act of divine creation, their different parts should appear "as if just off the production line"; in other words, these parts should be specifically designed to perform a given function efficiently. What we would not expect to find in this case is organs that appear to be a more or less lucky modification of others which perform a different function in other organisms. In other words, we would expect each living being to have its own parts, perfectly tailored to carrying out the functions with which they are entrusted.

Natural selection does not plan evolutionary change; it simply selects from among what is there. That is to say, it preserves those variations in existing organs which confer some advantage on individuals. Thus an organ can be modified and end up performing a function different from its original one. In this process, the organ in question may lose efficiency in performing its original task, provided that this loss is compensated by the advantage conferred by the new function.

But let us return to the theme of the human voice. When Wallace and Darwin argued over the nature of natural selection and its role in the origin of human beings, neither the anatomical basis nor the physiological mechanisms of speech were known. Today we understand that this human quality is based on the low position of our larynx, which in its turn is due to a modification of the structure of the upper airways common in other mammals. We also know that this particular position of the larynx restricts our capacity to drink and breathe at the same time, and is responsible for the unpleasant and dangerous phenomenon of choking. However, we consider these to be minimal inconveniences compared with the great advantage conferred by possessing a mechanism which allows us to modulate the sounds on which our language is based; there is no doubt that we have made a very good bargain here. Thus, we can recognize the mark of natural selection, and the trail of the evolutionary history of our species, in the anatomy of our phonetic apparatus.

Darwin can rest easy at Newton's side: once again, he was right.

17

The Meaning of Evolution

Until tonight, you thought that life was absurd. From now on you will know that it is mysterious.

Eric-Emmanuel Schmitt, *The Visitor*

The Action Replay of Life

In his book *Wonderful Life*, Stephen Jay Gould explains that we are not the inevitable result of evolution, but rather a mere circumstance of it – that if the videotape of life was rewound and played again from the beginning, planet Earth would now be populated by a completely different range of life forms, among which humans would be nowhere to be found.

It is, of course, impossible to perform this experiment by returning to the beginning of life. However, to a certain extent it has been conducted by natural means. For example, the platyrrhine monkeys in America did not evolve into forms of intelligence comparable to our own. Clearly, they were not subject to any "impulse" driving them toward "progress" or "perfection" (the same could be said of marsupials in Australia, and other similar cases of evolution in conditions of geographical isolation).

In any case, there is another way we can play Gould's game. If we assume that evolution is directed or tends spontaneously toward increasingly "higher" or more complex forms of life, we would expect the fossil record to reflect a history in which progressively more complex forms of life replace others, owing to their clear superiority, until humans appear.

255

But this analysis runs up against the problem of how to measure complexity. We need to be able to establish a scale which could be applied to living species or fossils, so that we can say: this species is at complexity level 3 or level 7 (the reader perhaps imagines that our species would be the only one at level 10). We have to admit that this is a difficult problem to solve.

In 1949 George Gaylord Simpson, one of the greatest paleontologists of the 20th century, published a very influential book entitled *The Meaning of Evolution*, in which he concluded that evolution had no purpose. Among other themes, in a chapter headed "Progress in Evolution" Simpson analyzed the question of whether there has been an increase in complexity over the course of the Story of Life. It was clear to him that when multicellular organisms (those with many cells) appeared, they represented an increase in complexity over the first, unicellular (single-celled) forms of life. According to Simpson, a second step toward greater complexity occurred when the major different types of multicellular organisms emerged (the technical term for these types is *phylum* – plural *phyla*); however, this progress took many directions, rather than following a single, privileged line. From this point on, it is impossible to compare the level of complexity within each of the lines. Simpson, who specialized in vertebrate paleontology, wrote that it would be a brave man who would attempt to prove that a human being is more complex than an ostracoderm (a kind of fishlike aquatic vertebrate which appeared more than 4 million years ago).

Meanwhile, we still have not defined what complexity is – no easy task (you might also like to try). One modern way of doing it would be to use the concept of complexity as it is applied to systems. A *system* is a combination of elements which interact with one another, giving rise to the properties of the system, and the more separate elements it has, the more different possibilities for interaction exist, rendering the system richer in functions, or more complex – in the sense of less predictable, less rigid, more variable, and also more adaptable.

Multicellular organisms are *self-regulated systems* composed of differentiated cells which form tissues and organs; these in their turn are organized into systems – respiratory, digestive, reproductive, excretory, circulatory, or nervous. The concept of the complexity of systems can be used to compare organisms in very different groups, for example to compare mammals with sponges or jellyfish: the latter are clearly much simpler forms of organization than mammals, with fewer differentiated elements, and we can consider them biological systems with a relatively low level of complexity. Who, however, would venture to assess the relative complexity of a bat and a lion?

Even if we compare organisms in terms of the number of genes expressed (the number which translate into proteins), we find that protozoa have more than bacteria, that invertebrates have more than protozoa, and vertebrates more than invertebrates. However, among vertebrates it is impossible to establish categories.

The issue of complexity is thus a real Gordian knot, and there is only one way to undo it: by cutting right through it. We can start from the basis that our species is, by definition, the most complex of all. However, if we compare ourselves with other primates, or other mammals, we have to ask what our greater complexity consists of. Only one of our systems, the central nervous system, can be considered more complex, and it would have to be this which ultimately won us top ranking on the scale of complexity. Following this line of reasoning, which sets humans as the measure of all things, it is clear that the closer a species is to *Homo sapiens* in its evolution, the closer its appearance and relationship to us, the greater will be its degree of complexity. Thus, mammals would be the most complex animals in the history of life; the most complex mammals would be primates, and the most complex of the primates, gorillas and chimpanzees, from which we are differentiated by only about one percent of our genes.

For the time being we shall not challenge this dubious system of measuring the degree of complexity of living beings using our species as a model. Let it be so, if the reader so wishes. What certainly is debatable is whether this supposed greater complexity constitutes an evolutionary advantage leading toward the triumph (progressive, linear, and inexorable) of the most complex. This would mean that the more complex would always win out in the struggle for existence over simpler organisms – right up to the arrival of the most complex being of all, the human being. Let us look at what the fossil record says about this.

Mammals constitute the group of vertebrates to which we belong, and this group is therefore universally considered the most "advanced" of all, far ahead of amphibians, reptiles, or birds. We might therefore imagine that since the time mammals appeared they have steadily replaced other terrestrial vertebrates. Many people have the vague idea that mammals emerged in a world dominated by dinosaurs, which they finally defeated thanks to their superiority.

However, the fossil record indicates quite the contrary. The direct ancestors of mammals, a group of reptiles known as *Therapsida* (so similar to the true mammals which came after that they have been called "mammaloid reptiles"), dominated the terrestrial ecosystems at the start of the Mesozoic (the Secondary era). This is normal, one might think – they were superior to other reptiles. However, at a certain point, roughly

200 million years ago, Therapsida began to decline, and were gradually replaced in terrestrial ecosystems by dinosaurs (birds later evolved from one particular group of dinosaurs, making them the true "living dinosaurs").

If the term "superior" can be applied in evolutionary biology to any particular group, in this case it would have to be applied to the dinosaurs. The Therapsida eventually became extinct, and although some evolved into the first mammals, these latter not only failed to win out over reptiles, but led a very discreet life throughout the remainder of the Mesozoic (without exception, all mammals were small in size). In fact it was the impact of a meteorite, not the superiority of mammals, which finished off the dinosaurs 65 million years ago. If it had not been for this "providential" meteorite, the evolution of terrestrial vertebrates would no doubt have been very different. (Some authors believe that the extinction of the dinosaurs may have been due to the atmospheric effects of a series of great volcanic eruptions; the important thing for our argument is that the disappearance of the dinosaurs had a nonbiological cause, and it is irrelevant whether this was a meteorite, a volcanic phenomenon, or any other geological disaster.)

Let us now turn to another case from the pages of the fossil record, this time even closer to ourselves. As we have noted in this book, within the order of primates we belong to a group, the hominoids, which includes a series of species, apes, with which we share many traits and, in truth, many genes. In fact, apes are also the primates with the most developed brains. Since they are the mammals most resembling us, it might be expected that their superiority led them to win out at least over the other monkeys existing at the time when they appeared. As we have noted, hominoids originated in Africa at least 23 million years ago and, from the moment when these primates made their way out of Africa to populate Europe and Asia (about 17 million years ago), they became the group of primates with the greatest evolutionary success, diversifying into a large number of species inhabiting the wide belt of forest which extended throughout much of the Old World. This might seem logical, since these were the most intelligent primates and their success prefigured the glorious future awaiting human beings.

Once more the fossil record tells us the opposite of what appears "logical." Between about 8 and 7 million years ago, hominoids ceased being the "kings of creation." A great ecological change caused their paradise to disappear. Astronomical factors, movements of continental landmasses, and the emergence of mountain ranges changed the climate and the composition of the atmosphere, resulting in deterioration of their habitat over much of its once enormous expanse. The forest gave way to more open ecosystems.

But it was not only that the change in vegetation reduced their living space; other primates, the *Cercopithecidae*, became more numerous and more varied than them. In military terms, the apes beat a retreat. Today they remain reduced to two species of chimpanzee and the gorilla in Africa, the orangutan in Sumatra and Borneo, and the gibbons of continental and island Southeast Asia. It is indeed significant that the little gibbon, the hominoid least like humans, is the most successful in terms of diversity and numbers. Despite pressure from humans, several million gibbons, of nine different species, still survive.

But about 5 or 6 million years ago, in some place in Africa, possibly the easternmost part, a specific type of hominoid began to emerge: the first hominid, our oldest ancestor. To begin with it differed very little from the ancestors of modern chimpanzees and gorillas. It could be considered the East African version of the same group. Later, at least 4 million years ago, this type of hominoid had developed a singular characteristic, never before seen. This was not a hominoid more intelligent than the others. No – it was a bipedal hominoid.

As time went on, bipedal hominoids adapted to the increasingly arid ecosystems covering much of Africa. They developed specialized dentition to cope with this. We have already seen that it is difficult to measure the intelligence of fossil species (even of living species), but the index of encephalization suggests that the most encephalized species 3 million years ago were not hominids but dolphins.

By about 2.5 million years ago the hominids had split into two different evolutionary lines. One of these, *Paranthropus*, developed a hypertrophied (overdeveloped) masticatory apparatus. The other line was that of the first representatives of the genus *Homo*, the first humans, who had a somewhat larger brain. Only from this moment did hominids become unique among living beings for their greater cerebral complexity. *Paranthropus* later became extinct, and subsequent humans modified their body structure, increased their brain size, and improved their technology. But even since the point when *intelligence* appeared in the biosphere, human evolution has not followed one single path, a straight line leading to ourselves. On the contrary, until a few thousand years ago a number of intelligent human species existed on earth. The fact that ours is the only one now in existence gives us the false perspective that it has always been so, that our ancestors succeeded one another in an ordered sequence, on a staircase we climbed step by step.

In short, neither the evolutionary history of mammals nor that of hominoids reflects a pattern whereby these groups appeared and progressively came to dominate other creatures thanks to their superior characteristics, particularly their intelligence. On the contrary, the fossil record in both cases shows a history of appearance and subsequent diversification, followed by almost complete extinction and ultimate resurgence. In the case of mammals, their resurgence was due to a favorable event of extraplanetary origin (or some geological disaster). In the case of hominoids, the resurgence was only partial, and resulted from the adaptation of one of their forms, the hominids, to a way of life completely new for primates, living in open environments: cerebral complexity had nothing to do with this adaptation.

What does all of this mean? To put it simply, were it not for a series of events unrelated to biology, such as a meteorite hitting the earth, the formation of mountain ranges, great continental shifts, and other smaller movements, we would not be here now to ponder these questions.

In other words, an extraterrestrial biologist at the beginning of the Mesozoic might have predicted great evolutionary success for the "mammaloid reptiles" and their descendants, and would have been mistaken (though incidentally, on this occasion the defeat of the mammaloid reptiles occurred without the need for any great disaster: the dinosaurs "fought clean" and defeated our ancestors purely through ecological competition).

Another alien biologist witnessing life on earth a few million years later might have predicted a great future for the dinosaurs, and would also have been mistaken. A third visitor, perhaps 10 million years ago, might have said that hominoids would reign for ever in the forests of the Old World, and would have been completely wrong.

The visitor from outer space who arrived 6 million years ago might be convinced that the entire group of hominoids faced imminent extinction. How could this extraterrestrial biologist have known that the ecological change which had such a detrimental effect on hominoids overall would favor the appearance of a type of bipedal hominoid which was later to give rise to a species – our own – which would populate the world and end up itself producing biologists? Even as little as 60,000 years ago, when the Neanderthals had spread throughout Europe, Central Asia, and the Middle East, who could have predicted that modern humans, our ancestors, would leave the African continent and be the cause of the extinction of the Neanderthals a few thousand years later? And now that we are beginning to know how things happened in the past, who would venture to make predictions about the future of the biosphere?

But the really significant issue is not the possibility of imagining what the future will be like – that is merely amusing speculation, a side issue. The important thing is that our capacity to predict the future is the measure of our knowledge of how the evolutionary process functions. But does this knowledge really depend only on us? If evolution followed specific tendencies or trajectories over time we could, by extrapolating these into the future, predict its course. Since the only trend which evolution appears to follow is that of adaptation in many different ways to changing environmental circumstances, the question of where species are going must remain unanswered.

This unpredictability of evolution indicates that nothing is decided beforehand, that anything is possible. It shows that the most flourishing biological group can become extinct because of changes in the physical environment or through competition with other groups of organisms. No form of life can be considered superior to others, because none is safe from mass extinction.

But if evolution is unpredictable, does this mean that it is governed by blind fate, that there are no laws, that everything is chaos and nothing can be explained? Is it reasonable to accept that disorder (or a lack of order) has produced such marvels of biology? Can random noise create a symphony just by chance?

The very core of evolution is pure chaos. Natural selection operates on genetic variants which arise without any relation to the activities or needs of the organisms. Mutation, which generates variation, is a *stochastic* process (regulated by chance). However, once a variant has arisen, whether it survives and spreads, or is eliminated and disappears, does not depend on chance. In the complex interrelation an organism maintains with other organisms and with the physical environment, certain variants offer those who carry them a greater capacity for survival and reproduction, while others reduce this capacity. Only the former will be called to survive. Natural selection is a deterministic process.

But in the longer term, on the scale of species and groups rather than that of individual organisms, does chance reign, or are there laws? According to traditional physics, including both Newtonian and the more modern quantum and relativist physics, total knowledge guarantees certainty, and unpredictability is due purely to our lack of knowledge. However, modern *chaos theory* predicts that there may be order, laws which we can know, within a dynamic system, but that its future behavior can nevertheless still be unpredictable. How can we understand this apparent paradox?

Organization and Chaos

Today we fear that the agricultural and industrial activities of the human race could destroy the ecological balance, causing the widespread extinction of species, however well adapted they may be. The introduction of animals, plants, or other organisms from one region into another (for example, bringing rabbits to Australia) can also have catastrophic effects on the ecological balance. We are likewise concerned about the effects on the biosphere of climate change caused by the emission of greenhouse gases such as carbon dioxide, and by the destruction of the ozone layer. We are all increasingly aware that ecosystems, which are always in fragile equilibrium, are composed of many species with a long history.

But all of these things about which we are (with reason) concerned have occurred many times in the past, through natural causes. The continents and oceans were not always distributed over the earth's surface as they are now, and the different regions where life is found have separated and come together in many different ways, giving rise to various exchanges between species in which some were favored while others lost out. Climate change is a frequent occurrence in the Story of Life, and has a powerful impact on ecosystems (the alternation between ice ages and interglacial periods such as the one in which we are now living – which will not last forever – is but one example). The composition of the atmosphere has also varied.

Finally, even if the physical environment remained unchanged, the appearance, through evolution, of new species introduces a fundamental instability factor into ecosystems, and means that they are always changing. Organisms show adaptations, developed over the period of their evolution, which only make sense in relation to the ecological niches occupied by their species. In a community, all populations are related to one another in intricate networks, through which energy and matter flow. To take the example of a primate, any change in the plants on which it feeds or the animals it eats, the predators which prey on it, its competitors, or its parasites, will have unpredictable consequences for the survival of the species, which will have to evolve in its turn, adapting to its new circumstances.

In other words, the biosphere is an enormously complex macrosystem, made up of many different elements which organize themselves according to a hierarchy of levels (cell, tissue, organ, system, individual, family group, social group, population, community, ecosystem), and interact in many different ways at all levels. This kind of system makes it extremely difficult to identify the basic laws according to which it functions, even when the constituent elements remain the same. However, to complicate things further, the evolution of species means that the biosphere is an unstable system, far from equilibrium, which has never remained static and whose composition (species, and therefore their various interactions) has changed over time.

It follows from all of this that the evolution of species is subject to the influence of so many factors that in practice it is impossible to predict its future. This does not mean that evolution depends on pure chance (in the popular sense of incomprehensible chaos). On the contrary, it can be understood, even if only with hindsight, like the weather. To a certain extent, the well-known example of chaos theory called the "butterfly effect" can be applied here: the flutter of a butterfly's wings in Beijing can cause rain to fall in New York (to say nothing of the case where the delicate movement of an insect's wings is replaced by a meteorite many miles in diameter traveling at full speed toward earth).

But is this a merely technical question? Is our uncertainty due only to the complexity of the problem? Chaos theory goes further, asserting that even if we knew all these factors and interactions in full detail, the future cannot be known, simply because it is not predetermined. This is the end of certainty, which is replaced by probabilities.

In his book *On Aggression*, Konrad Lorenz recounts how Alfred Kühn ended a lecture with Goethe's words: "It is the greatest joy of the man of thought to have explored the explorable and then calmly to revere the unexplorable." As the audience broke into applause, Kühn raised his voice and cried: "No, not *calmly*, gentlemen; never *calmly*!" Should we assume that at this point investigation into the nature of evolution is closed? Has everything already been said?

We believe, on the contrary, that there is still much work to be done. Newtonian physics speaks of trajectories which can be expressed in terms of equations. If we know the initial conditions, these trajectories are predictable and reversible, like a pendulum – first here, then there. In these equations time does not exist: it is merely an illusion in which the future and the past meet. Quantum physics simply replaces trajectories with wave functions, but the symmetry with regard to time does not change. Biological evolution, in contrast, is an irreversible process which unfolds over time, surprising us at each moment, and follows no trajectory (or trend). How can we reconcile physics and biology? If chaos theory is correct, there is, as Ilya Prigogine (winner of the 1977 Nobel Prize for Chemistry) says in his book *The End of Certainty*, a narrow path between two equally alienating conceptions of the world. Either we live in a deterministic world governed by immutable laws which leave no place for novelty (and where the greatest novelty of all, evolution, would be impossible), or we are in "a world...where everything is absurd, acausal and incomprehensible," subject to pure chance. It is the task of the Darwins of the present and the future to travel this narrow path.

Epilogue

Living organisms had existed on earth, without ever knowing why, for over three thousand million years before the truth finally dawned on one of them. His name was Charles Darwin.

Richard Dawkins, *The Selfish Gene*

The Never-ending Story

In his *Dictionary of the Cinema*, Fernando Trueba tells how the famous French film director François Truffaut declared that when he went to the cinema and saw a group of characters digging a tunnel for an hour and a half and at the end of the film the tunnel was no use, he believed he should get his money back. Although we cannot give our readers their money back, we would not like to leave them with the unpleasant impression that more than 3,000 million years of evolution have been completely pointless, and that we are just a species like any other. Because in fact this is not true.

In his well-known book *The Meaning of Evolution*, which we referred to in a previous chapter, George Gaylord Simpson explains that evolution has no aim, but then rejects the suggestion that this means that man is "just an animal." To begin with, Simpson does not see any animal species as "just an animal," because each one has its own peculiarities, shared with no other species. But ours is exceptional in many very important aspects. We are the most intelligent species, and this intelligence (however defined) has enabled us to occupy all the different landscapes of the planet

264

and to develop a technology. Whether this technology is used for good or ill, for healing or for killing, we are capable of drastically altering our environment and even the entire biosphere.

Eörs Szathmáry and John Maynard Smith point to a series of great transitions in the Story of Life. Taking a look at these stellar moments of evolution can help us to evaluate the true significance of our appearance in the biosphere.

The first transition consisted of the transition from *free molecules*, or "replicants," capable of replicating themselves, into populations of "replicants" enclosed in a single container.

The second great transition consists of the association of "replicants" in chromosomes.

The third was the change in the composition of the *inheritance molecule*, which changes from being RNA (ribonucleic acid) to DNA (deoxyribonucleic acid). DNA contains the genetic information, while RNA functions as an intermediary in the synthesis of proteins. With the appearance of DNA comes the genetic code shared by all living beings (so in fact these three steps are hypothetical, and actually form part of the Prehistory of Life as we know it today).

The fourth transition moves from the organisms known as *prokaryotes* (bacteria and blue-green algae or cyanobacteria) to the first *eukaryotic* organisms. This is the category to which we belong, and it is characterized by cells which have a nucleus and a series of organelles such as chloroplasts and mitochondria. Many authors believe that the organelles are former prokaryotes which have become integrated into the eukaryotic cell; mitochondria, for example, contain small circular molecules of DNA, like those of bacteria.

The fifth great transition occurred when organisms moved from asexual reproduction ("self-cloning") to sexual reproduction.

The sixth transition, the next step, led from the organisms of the *kingdom of Protista* (all unicellular and eukaryotic organisms) to *multicellular* organisms, made up of many cells (these are also eukaryotes). This step is thought to have occurred at least three times, independently, giving rise to the animal and plant kingdoms and the *kingdom of fungi*. (In addition to these four kingdoms there is also the *Prokaryota kingdom*, mentioned above; some divide this in turn into two further kingdoms, that of the "normal" bacteria, or *Eubacteria*, and a kingdom of a few prokaryotes which live in extreme conditions and form the *kingdom of Archaebacteria*). These six great transitions occurred at different moments, but all far back in time, since multicellular organisms have existed for 680 million years.

The seventh transition marked the shift from isolated organisms to colonial organisms which include categories of non-reproducing organisms, like the caste societies of insects.

Many of these transitions have some very interesting features in common. Essentially, we see a pattern where elements which live and reproduce in isolation lose some of their independence and come together to form larger entities (and can no longer reproduce alone): the replicant becomes integrated into the first container and then into the chromosome; the ancient, free-living prokaryotes become associated in the eukaryotic cell; the protist becomes a cell in a multicellular organism; the individual has to live in a colony, in order to ensure the survival of genes shared by the individuals; with sexual reproduction all organisms depend on others to continue their line, and therefore need to belong to a population.

Moreover, several of these transitions are also marked by specialization and division of labor among the elements which have come together: the different genes code for different proteins; the cellular organelles have different functions; the different cells of a multicellular organism form tissues of very different kinds; each caste has its own function in the colony.

Since we do not belong to a social insect species, we might take the view that the most important events in the process of evolution which leads to ourselves had already taken place at least 680 million years ago, and that since then nothing of any major significance has happened. According to Eörs Szathmáry and John Maynard Smith, nothing could be further from the truth.

The eighth great transition took place a very short time ago, and consisted in the shift from primate societies to human societies, with the appearance of *articulated language* – a unique, revolutionary, and extremely powerful system for transmission of information (which has made it possible, among other things, to write this book using an alphabet of 26 letters). It has, after all, been worth the long wait. Although genetically we are primates very close to chimpanzees, and a product of evolution, we represent a type of organism radically different from all the others. We are the only beings which question the meaning of our own existence.

But let us not get carried away now by excessive triumphalism, because we must also recognize that since the beginning of scientific thinking among the ancient Greeks, there have been concerted efforts to position our species with its back to nature, or even worse, above nature. This is the source of some of the great problems afflicting humanity in the present.

Only since Darwin has it been understood that we are not the *chosen species* but, as Robert Foley suggests, a *unique species* among many other unique species – albeit an amazingly intelligent one.

And it remains a paradox that so many centuries of science have led us to know what any Kalahari Bushman, any Australian Aborigine, or any of our ancestors who painted bison in the caves of Altamira knew full well: that it is not the earth that belongs to man, but man who belongs to the earth.

Bibliography

In addition to the books cited in the text, we felt it would be useful to provide a brief list of some of the most important recent publications which might be of interest to the reader who wants to study human evolution in more depth. Inevitably some important works will have been omitted, although we have attempted to ensure that, though not all the references available are listed here, all those that are listed are available.

Books

Aguirre, E., ed.
   1988 Paleontología Humana. Barcelona: Prensa Científica.
Aiello, L., and C. Dean
   1990 An Introduction to Human Evolutionary Anatomy. San Diego: Academic Press.
Ayala, F. J.
   1980 Origen y evolución del hombre. Madrid: Alianza.
Bermúdez de Castro, J. M., J. L. Arsuaga, and E. Carbonell, eds.
   1992 Evolución humana en Europa y los yacimientos de la sierra de Atapuerca. Valladolid: Junta de Castilla y León, Consejería de Cultura y Turismo.
Bertranpetit, J., ed.
   1993 Orígenes del hombre moderno. Barcelona: Prensa Científica.
Bowler, P.
   1986 Theories of Human Evolution. Oxford: Blackwell.
Brain, C.
   1981 The Hunters or the Hunted? Chicago: University of Chicago Press.

Conroy, G.
1990 Primate Evolution. New York: W. W. Norton.
Day, M.
1986 Guide to Fossil Man. London: Cassell.
Fleagle, J.
1988 Primate Adaptation and Evolution. San Diego: Academic Press.
Foley, Robert
1987 Another Unique Species. Harlow: Longman.
Fossey, Dian
1983 Gorillas in the Mist. Boston: Houghton Mifflin.
Goodall, Jane
1971 In the Shadow of Man. Boston: Houghton Mifflin.
1990 Through a Window. Boston: Houghton Mifflin.
Johansen, D., and M. Edey
1981 Lucy: The Beginnings of Humankind. New York: Simon and Schuster.
Johansen, D., and B. Edgar
1996 From Lucy to Language. New York: Simon and Schuster.
Jones, S., R. Martin, and D. Pilbeam, eds.
1992 The Cambridge Encyclopedia of Human Evolution. Cambridge: Cambridge University Press.
Klein, R.
1989 The Human Career. Chicago: University of Chicago Press.
Landau, M.
1991 Narratives of Human Evolution. New Haven: Yale University Press.
Leakey, Richard
1981 The Making of Mankind. New York: E. P. Dutton.
Leakey, Richard, and R. Lewin
1992 Origins Reconsidered. New York: Doubleday.
Le Gros Clark, Wilfrid
1969 The Antecedents of Man. Chicago: Quadrangle.
Lewin, R.
1987 Bones of Contention: Controversies in the Search for Human Origins. New York: Simon and Schuster.
Martin, R.
1990 Primate Origins and Evolution. London: Chapman and Hall.
Napier, J., and P. Napier
1985 The Natural History of the Primates. London: British Museum (Natural History).
Reader, J.
1981 Missing Links. Harmondsworth: Penguin.
Rightmire, G.
1990 The Evolution of *Homo erectus*. Cambridge: Cambridge University Press.

Savage-Rumbaugh, S., and R. Lewin
1994 From Kanzi: The Ape at the Brink of the Human Mind. New York: John Wiley.
Schultz, A.
1969 The Life of Primates. New York: Universe Books.
Stringer, Christopher, and Clive Gamble
1993 In Search of the Neanderthals: Solving the Puzzle of Human Origins. London: Thames and Hudson.
Stringer, Christopher, and Robin McKie
1996 African Exodus. London: Jonathan Cape.
Szalay, F., and E. Delson
1979 Evolutionary History of the Primates. New York: Academic Press.
Tattersall, Ian
1995 The Fossil Trail. New York: Oxford University Press.
1995 The Last Neanderthal. New York: Macmillan.
Tattersall, Ian, E. Delson, and J. Van Couvering, eds.
1988 Encyclopedia of Human Evolution and Prehistory. New York: Garland.
Trinkaus, E., and P. Shipman
1992 The Neanderthals. New York: Vintage.
Walker, A., and Richard Leakey, eds.
1993 The Nariokotome *Homo erectus* Skeleton. Berlin: Springer-Verlag.
Walker, A., and P. Shipman
1996 The Wisdom of the Bones. New York: Alfred A. Knopf.
Wolpoff, M.
1980 Paleoanthropology. New York: Alfred A. Knopf.
Wood, B.
1991 Koobi Fora Research Project, vol. 4. Hominid Cranial Remains. Oxford: Clarendon Press.

Articles

It is not our intention to give an exhaustive list of scientific articles, which are generally published in specialist journals not easily accessible to the majority of readers. However, we refer directly to some authors and papers in our text, and it seemed right to us to cite the sources.

Aiello, L., and R. Dunbar
1992 Neocortex Size, Group Size, and the Evolution of Language. Current Anthropology 34:184–193.
Aiello, L., and P. Wheeler
1994 Brains and Guts in Human and Primate Evolution: The Expensive Organ Hypothesis. Current Anthropology 36:199–221.

Arensburg, B., et al.
1990 A Reappraisal of the Anatomical Basis for Speech in Middle Palaeolithic Hominids. American Journal of Physical Anthropology 83:137–146.

Arsuaga, J. L., et al.
1997 Size Variation in Middle Pleistocene Humans. Science 277:1086–1088.

Arsuaga, J. L., and I. Martínez
1989 Paleontología humana: el origen de la humanidad. *In* Paleontología. E. Aguirre, ed. Madrid: CSIC.

Arsuaga, J. L., et al.
1993 Three New Human Skulls from the Sima de los Huesos Middle Pleistocene Site in Sierra de Atapuerca, Spain. Nature 362:534–537.

Berge, C.
1900 Interprétation fonctionnelle des dimensions de la cavité pelvienne de *Australopithecus afarensis* (AL 288-1). Zeitschrift für Morphologie und Anthropologie 78:321–330.

Bermúdez de Castro, J. M., et al.
1997 A Hominid from the Lower Pleistocene of Atapuerca: Possible Ancestor to Neanderthals and Modern Humans. Science 276:1392–1395.

Bromage, T., and M. Dean
1985 Reevaluation of the Age at Death of Plio-Pleistocene Fossil Hominids. Nature 317:525–528.

Brunet, M., et al.
1995 *Australopithecus bahrelghazali*, une nouvelle espèce d'hominidé ancien de la région de Koro Toro (Tchad). Comptes Rendus de l'Académie des Sciences de Paris, série IIA 322:907–913.

Cabrera, V., and J. Bischoff
1989 The Dates for Upper Paleolithic (Basal Aurignacian) at El Castillo Cave (Spain). Journal of Archaeological Science 16:577–584.

Cann, R., M. Stoneking, and A. Wilson
1987 Mitochondrial DNA and Human Evolution. Nature 325:31–36.

Carbonell, E., et al.
1996 Lower Pleistocene Fossil Hominids and Artefacts from Atapuerca-TD6 (Spain). Science 269:826–830.

Cavalli-Sforza, L.
1991 Genes, Peoples and Languages. Scientific American 265:72–79.

Cerling, T., et al.
1997 Global Vegetation Change through the Miocene/Pliocene Boundary. Nature 389:153–158.

Coppens, Y.
1994 East Side Story: The Origin of Humankind. Scientific American 270: 62–69.

deMenocal, P.
1995 Plio-Pleistocene African Climate. Science 270:53–58.

Falk, D.
1993 Evolution of the Brain and Cognition in Hominids. James Arthur Lectures on the Evolution of the Human Brain. New York: American Museum of Natural History.

Fernández-Jalvo, Y., et al.
1996 Evidence of Early Cannibalism. Science 271:269–270.

Foley, R., and M. Lahr
1998 Mode 3 Technologies and the Evolution of Modern Humans. Cambridge Archaeological Journal 7:3–36.

Gagneux, P., D. Woodruff, and C. Boesch
1997 Furtive Mating in Female Chimpanzees. Nature 387:358–359.

Hammer, M., and S. Zegura
1996 The Role of the Y chromosome in Human Evolutionary Studies. Evolutionary Anthropology 5:116–134.

Harcourt, A.
1994 Sexual Selection and Sperm Competition in Primates: What are Male Genitalia Good For? Evolutionary Anthropology 4:121–129.

Hublin, J. J., et al.
1995 The Mousterian Site of Zafarraya (Andalucia, Spain): Dating and Implications on the Palaeolithic Peopling Processes of Western Europe. Comptes Rendus de l'Académie des Sciences de Paris, série II 321(10):931–937.
1996 A Late Neanderthal Associated with Upper Palaeolithic Artefacts. Nature 381:224–226.

Johanson, D., et al.
1986 New partial skeleton of *Homo habilis* from Olduvai Gorge, Tanzania. Nature 327:205–209.

Kappelman, J.
1997 They Might be Giants. Nature 387:126–127.

Kimbel, W., D. Johanson, and Y. Rak
1996 Systematic Assessment of a Maxilla of *Homo* from Hadar, Ethiopia. American Journal of Physical Anthropology 103:235–262.

Klein, J., N. Takahata, and F. J. Ayala
1993 Mhc Polymorphism and Human Origins. Scientific American 269:78–83.

Krings, M., et al.
1997 Neanderthal DNA Sequences and the Origin of Modern Humans. Cell 90:1–20.

Leakey, M. G., et al.
1995 New Four-million-year-old Hominid Species from Kanapoi and Allia Bay, Kenya. Nature 376:565–571.

Lieberman, P., et al.
1992 The Anatomy, Physiology, Acoustics and Perception of Speech: Essential Elements in the Analysis of the Evolution of Human Speech. Journal of Human Evolution 23:447–467.

Lovejoy, C.
1981 The Origin of Man. Science 211:341–350.

Manzi, G., A. Vienna, and G. Hauser
1996 Developmental Stress and Cranial Hypostasis by Epigenetic Trait Occurrence and Distribution: An Exploratory Study on the Italian Neanderthals. Journal of Human Evolution 3:511–527.

McHenry, H.
1992 How Big were Early Hominids? Evolutionary Anthropology 1:15–20.

Moyà-Solà, S., and M. Köhler
1997 A Dryopithecus Skeleton and the Origins of Great-ape Locomotion. Nature 379:156–159.

Parés, J. M., and A. Pérez-González
1995 Paleomagnetic Age for Hominid Fossils at Atapuerca Archaeological Site, Spain. Science 269:830–832.

Plavcan, J., and C. Van Schaik
1996 Intrasexual Competition and Body Weight Dimorphism in Anthropoid Primates. American Journal of Physical Anthropology 103:37–68.

Rosenberg, K., and Trevathan, W.
1997 Bipedalism and Human Birth: The Obstetrical Dilemma Revisited. Evolutionary Anthropology 4:161–168.

Ruff, C., E. Trinkaus, and T. Holliday
1997 Body Mass and Encephalization in Pleistocene *Homo*. Nature 387: 173–176.

Schrenk, F., et al.
1993 Oldest *Homo* and Pliocene Biogeography of the Malawi Rift. Nature 365:833–836.

Semendeferi, K., et al.
1998 The Evolution of the Frontal Lobes: A Volumetric Analysis Based on Three-dimensional Reconstructions of Magnetic Resonance Scans of Human and Ape Brains. Journal of Human Evolution 32:375–388.

Seyfarth, R., and D. Cheney
1992 Meaning and Mind in Monkeys. Scientific American 267:122–129.

Suwa, G., et al.
1993 The First Skull of *Australopithecus boisei*. Nature 389:489–492.

Swisher III, C., et al.
1994 Age of the Earliest Known Hominids in Java, Indonesia. Science 263:1118–1121.
1997 Latest *Homo erectus* of Java: Potential Contemporaneity with *Homo sapiens* in Southeast Asia. Science 274:1870–1874.

Szathmáry, E., and J. Smith
1995 The Major Evolutionary Transitions. Nature 374:227–232.

Tague, R., and C. Lovejoy
1986 The Obstetric Pelvis of A.L. 288–1 (Lucy). Journal of Human Evolution 15:237–255.

Bibliography

Vrba, E.
1987 Late Pliocene Climatic Events and Hominid Evolution. *In* Evolutionary History of the "Robust" Australopithecines. F. Grine, ed. New York: Aldine de Gruyter.

Wheeler, P.
1993 Human Ancestors Walked Tall, Stayed Cool. Natural History 102:65–67.

White, T., G. Suwa, and B. Asfaw
1993 *Australopithecus ramidus*, a New Species of Early Hominid from Aramis, Ethiopia. Nature 371:306–312.

Index

Note: page numbers in italics refer to figures or illustrations

abductor muscles 66–7, 70–1
Aboriginal Australians 222–3, 267
El Abric Romani deposits 214
Acheulian industry 107, *109,* 185, 186, 206
Adapiformes 28, *30*
adaptation: evolution 261; fossils 58;
    intelligence 169; Lamarck 14; natural
    selection 10, 238; primates 19
adaptive convergence 34, 110–11
Aegyptopithecus 31
Africa 22, 27, 226; fossil sites *52, 88, 99,*
    174, 209; hominoids 32–3, 172;
    ocean influences 29, 43, 44; premodern
    224–5; *Proconsul* 30, *30*–2;
    rainfall 43–4; *see also individual*
    countries
Africa, East 43–4, 87, 114–15, 186
Africa, North 190
Afropithecus 30
agriculture 36, 130, 144, 261
Aguirre, Emiliano 184, 193
Aiello, Leslie 127, 142–4, 169–70
alcelaphins 91
alleles 50
Allen's rule 200
allometry 118–19
altruism 164, 251–2
America, North 27, 28, 29, 37
America, South 29
Amud site 221
ancestors 15, 19–20, 22–4, 109–11, 216
Ankarapithecus 32
Annaud, Jean-Jacques 216

antelopes 53, 87
Aotus trivirgatus 21
apes 22; anthropoid 21, 66, 169–70;
    birth 147, 150; diet 129–30, 131;
    sexuality 155; teeth 135–6; *see also*
    primates
Apidium 31
Arago cave 191, 209
Arambourg, Camille 89
Archaebacteria kingdom 265
Ardipithecus ramidus 51; forest-dwellers
    142, 172; teeth 53, 112–13, 136–7, 138,
    166
L'Ardreda deposits 214
Arensberg, Baruch 249
argon dating methods 60, 179
arm–leg length 24, 33, 79, 104, 107
Arsuaga, Juan Luis 184, 187, 195
art 212, 214–16, 218–19
Arvicola cantiana 185
Asfaw, Berhane 51
Asia 172, 179–82, 190
Atapuerca Mountains 191, 192, 196, 235,
    249
Atlantic Ocean 29, 43, 44
Auel, Jean M. 216
Aurignacian industry 212, 214
Australia 190, 222–3
australopithecines 2, 58; bipedalism 73–4;
    body mass 122; brain size 81;
    dentition 86; encephalization 123;
    environment 162; facial skeleton 81, *86;*
    hands 101; ilium 81; life

Index

australopithecines (*cont.*)
  stages 153; monogamy 163;
  pelvis 150; sexual dimorphism 173;
  vegetarian diet 142
Australopithecus afarensis: body mass 122;
  brain size 123, 151–2;
  encephalization 173; Hadar area 54;
  hands 141–2; jawbones 58–9; Laetoli
  site 75–6; mastication 113; neck 73;
  phalanges 79; sagittal/nuchal
  ridges *55,* 81; sexual dimorphism *84,
  85,* 165–6, *167, 168*; skeleton *80, 82*;
  teeth 59, 137–9, 166
Australopithecus africanus 61–2; brain
  size 123, 151–2; disappearance 87;
  gender difference *63,* 78; molars 146;
  sagittal ridges 81; South Africa 173;
  teeth 137–8, 139
Australopithecus anamensis 53, 54, 113,
  137–8, 172
Australopithecus bahrelghazali 58, 59, 113
Awash River site 51, 54, 60
Ayala, Francisco 236

baboons 18, 47, 135, 141, 161, 163, 166
Bañolas deposit 192
Bar-Yosef, Ofer 221
Baringo, Lake 98
Barroso, Cecilio 215
bears 134, 186–7, 192
Berge, Christine 150
Berger, Lee 141
Bergmann's rule 200
Bering Strait 27
Bermúdez de Castro, José María 184, 187,
  194–5
Biache-Saint-Vaast deposit 191, 208
Bilzingsleben deposit 191, 207, 209, 218
biomechanical efficiency 77
biosphere 262
bipedalism 3, 9, 62, 259; abductor
  muscles 70–1;
  australopithecines 73–4, 113;
  ecological niche 169;
  food-carrying 164; hominids 53, 54;
  legs 24; Lucy 1–2, 71, 76;
  monogamy 163, 166, *167, 168*;
  pelvis 24; savanna 76; skeleton 77–8;
  tools 65, 77
birth canal 147, *148, 149,* 150–2, 200
birth intervals 164
Bischoff, James 194, 214
Black Eve hypothesis 230–1
Bolomor molar 191
Bonis, Louis de 32
bonobos *20, 22,* 24, 101, 160

Border Cave 224
Boule, Pierre Marcelline 226
bovids 87, 91
Boxgrove site 182, 185, 191
brachiation *23,* 24, *25*
Brain, Bob 141
brain 116, *117,* 124, *125*; asymmetry 126,
  127; language 127, 241–3;
  lateralization 126, 127; lobes 124, *125,*
  126, 127; mass 123, 128, 152, 169;
  metabolism 142, 143
brain size 15, 62, 117–20; anthropoids 21;
  australopithecines 81; *Australopithecus
  afarensis* 123, 151–2; birth
  canal 151–2; body 118;
  chimpanzees 118, 119, 151–2; frontal
  lobe 127; *Homo* 102, 107, 113–14,
  143–4, 259; *Homo ergaster* 175;
  intelligence 127–8; longevity 153–4;
  Neanderthals 201; social behavior 169;
  stomach size 142–4
Branisella boliviana 29
Bräuer, Gunter 226
breastfeeding 243, *250*
Broca, Paul 128
Broca's area *117,* 125–6, 241, 242
Broken Hill 209, 247
Bromage, Tim 94, 98, 153
Broom, Robert 87, 89, 91
brow ridges: australopithecines 55, 81;
  Gran Dolina 188; Hahnöfersand
  fossil 215; *Homo* 102, *103,* 105, *106,*
  113–14, 180; Neanderthals 203
Brown, Peter 222
Brunet, Michel 58
Brunhes chron 185, 191
burins 212, *213*
burying the dead 206–7, 218
Byrne, Richard 170

Cabezo Gordo deposit 192
calcium 205
Callitrichinae 21
Can Llobateres deposit 32
Canidae 134
canines 132, 134–6, 137, 139, 166
Cann, Rebecca 230, 237
cannibalism 64, 187, 192
carbohydrates 130, 134
carbon-14 dating method 60
carbon dioxide 40, 44–8, 172, 262
Carbonell, Eudald 184, 186, 187
Caribbean Sea 29
carnivores 129–30, 132, 134, 142–4,
  186–7
Carretero, José Miguel 195

carrion 143, 186–7
El Castillo cave 214
catarrhines 21–2, 111, 145–6
catastrophe 12, 39
Cavalli-Sforza, Luigi 232
cave bear 134, 192
cave-dwellers 186–7
cave paintings 214–16, 219
cellulose 130, 134
Cenozoic 27
Ceprano site 191
Cercopithecidae 22, 47, 162, 172, 259
Cercopithecus aethiops 240, 241
cereals 130, 139, 144
cerebellum 116, *117*
cerebral cortex *117*, 124, 241, 242–3
cerebrum 116, *117*, 124
Cerling, Thure 46
cetaceans 120
Chad fossils 58–9
chaffinches 35–6
La Chaise-Abri Suard deposit 191, 208
chaos theory 261, 263
La Chapelle-aux-Saints 248
Chatelperronian industry 213–14, 216,
   217
Chauvet, Grotte 214–16
Chemeron site 51
Cheney, Dorothy 240
Chenjiawo site 181
childbirth 70, 146–7, *148, 149,* 150
children 62, 64, 197, 207
Chilo 171
chimpanzees, common *(Pan
   troglodytes)* 22, *23,* 49–50; airways *244;*
   birth 147, *148;* brain size 118, 119, 123,
   151–2; diet 130; distribution 162–3,
   217, 259; estrus 146, 155–6; extensor
   muscles *69;* femur 73; forest habitat 79;
   gluteus medius *67;* hamstring
   muscles *72;* hands 101;
   language 240–1; life expectancy 146;
   locomotion 26–7, 77–8;
   menstruation 165; pelvis 68; sexual
   dimorphism 159–60; skeleton *25;*
   social groups 159, 170, 251–2;
   teeth 138–9, 146; testicles 160–1;
   tools 98–100
chimpanzees, pygmy *(Pan paniscus)* 20, 22,
   160, 217, 259
China 102, 181, 185, 209, 224
Chinese medicine 34
chloroplasts 265
choking 243–4, 254
chopping tools 100
chordates 15

chromosomes 228–9, 231–2, 236, 265
chron 61, 185
clades 22
Clark, Sir Wilfrid Le Gros 65, 141, 152–3
Clarke, Arthur C. 49, 64
Clarke, Ron 141
clavicle 24, 111
cleavers 107, *109*
climate change: Africa, East 87; average
   temperature *37;* carbon dioxide 262;
   catastrophe 39; global 172;
   grasslands 143; habitats 87;
   intelligence 64–5; latitude 37, 44;
   Paranthropus 175
colobus monkey *25,* 51, 53, 135
comparative sociobiology 158–60
competition 47, 144, 251
complexity 256–7
Conrad, Joseph 145
continental drift 40, 44
Coon, Carleton 227
cooperation 251–2
Coppens, Yves 56, 57, 59, 89
copulation 150
Cosquer, Grotte 215
La Cotte de St. Brelade 207
coxal bones 69–70, 71, 191
cranial capacity 125, 196
Cretaceous 27
Crichton, Michael 233, 235
Cro-Magnon man 134, 215, 216–19
culture 158
Curtiss, Garniss 179

Dar-es-Soltan II cave 224
Dart, Raymond 61–2, 64, 141
Darwin, Charles 264, 267;
   Autobiography 9; *Beagle* trip 35–6; *The
   Descent of Man* 1, 197; evolution of
   species 3, 14, 110; natural
   selection 10–13, 254; *On the Origin of
   Species* 197; sexual selection 160;
   Wallace 10, 253–4
Darwinism 16, 171
Dawkins, Richard 164, 264
days/nights *41*
Dean, Chris 153
Dederiyeh site 221
deMenocal, Peter 44
dentition: australopithecines 59, 86,
   172; carnivores 134; catarrhines
   21–2; hominoids 32–4; life
   stages 145–6; *Oreopithecus* 34;
   Paranthropus 34, 59, 95, 143;
   primates 19–20; *see also* teeth
Descartes, René 116

diastema 135, 139
diet 129–30, 132, 139, 141, 169, 172; *see also* carnivores; herbivores; vegetarians
digestion 134, 142–3
Dinesen, Isak 226
dinosaurs 39, 258, 260
Dmanisi site 182
DNA (deoxyribonucleic acid): fossils 233–4; humans 234–5; inheritance 228–9, 265; mitochondrial 229–31, 232, 233–4; nuclear 231, 232
dolphins 110, 111, 120, 169, 259
drupes 131
Dryopithecus 32, *33*
Dryopithecus laietanus 33
Dubois, Eugène 179, 180
Dunbar, Robin 127, 169–70

earth–sun orbits 40–3
East Side Story hypothesis 44, 57–9
echinoderms 15
ecological factors 35–6, 46–8, 261–2
ecological niche 10, 76, 94, 160, 169, 172
education 158
Egypt 28, *31*; *see also* El Fayum
Ehringsdorf deposit 191, 208
Edelweiss group 192–3
Einstein, Albert 5
Eldredge, Niles 12
electron spin resonance 60–1
elephant 46, 118
enamel on teeth 53, 113, 136–7, 153
encephalization 119, 120; australopithecines 123; *Australopithecus afarensis* 173; hominids 117, 123–4, 259; humans 128; Neanderthals 201
end-scrapers 212, *213*
energy 132, 169
Engis site 197
environmental factors 11–12, 162
Eocene 28
epiglottis *244*
Equidae 46, 91
equinox 40, *41,* 42
Erythrocebus patas 18, 47
estrus 146, 155–6, 159, 165
Ethiopia 51
ethology 156, 157
Eubacteria kingdom 265
eukaryotes 265
euprimates 27, *29*
Eurasia 27, 28, 34, 37
Europe 27, 172, 182–6, 191, 211; *see also* *individual countries*

evolution 39–40, 251, 254, 260–1, 263
evolution of species: Darwin 3, 14, 110; directionality 14; divergence 15; Lamarck 9–10; natural selection 10–13; progress concept 13–16; Story of Life 15, 109, 256, 262, 265
evolutionary diagrams 29, *112, 114, 189, 210*
extension *69,* 71

facial characteristics: australopithecines 81, *86;* Gran Dolina Child 188; nasal bones 105, *106,* 188; Neanderthals *193, 202,* 203, *204,* 205, 208–9; Sangiran site 181; *see also* brow ridges; nuchal ridge
Falguères, Christophe 194, 223
Falk, Dean 125–6, 127
El Fayum deposits 28, *31*
Felidae 17, 132
femur 73, 79, 104, 107
La Ferassie 1 site 198
figurines 215, 218
fire 206–7
first family concept 56, 78
fission-track dating 60
Foley, Rob 162, 166, 209, 232, 267
footprints, Laetoli site 74–6
foramen magnum 73, *74,* 81, 247
foraminifera 38–9
forest habitat: chimpanzees 79; gorillas 79, *168;* hominids 51, 53; Miocene 57; primates 19; reduction 46–7, 48, 258
Fossey, Dian 158
fossils: dating 59–61; DNA 233–4; hominids 51, 53, 57–9, 172–3; pollen grains 39; sedimentary deposits 58, 59; sex of 195–6; *see also* paleontology
Frayer, David 248
Frisch, Karl von 156
frontal lobe 124, *125,* 126, 127
fruit 130, 131, 159
Fuente Nueva-3 186
funerary practice 194, 206–7, 218

Gagneux, Paul 159
Galápagos Islands 35–6
Galdikas, Biruté 158
Gamble, Clive 207–8, 248
Garrod, Dorothy 221
geese 156
geladas *(Theropithecus gelada)* 18, 47, 135, 141, 163, 166
gender difference 78, 164–5; *see also* sexual dimorphism

genes 228; behavior 157;
  chromosomes 236; difference 49–50,
  216–17; flow 227; natural
  selection 156, 164, 261; neutral/
  non-neutral 50
genetic defects 238
genetic manipulation 238
geochronology 61, 179, 194
Gibart, Josep 182
gibbons *(Hylobates lar)* 22, *23*, 24, 161;
  locomotion 26; monogamy 163;
  species 217, 259
Gibraltar, Straits of 189, 190
Gibraltar site 197
Gigantopithecus 32
Gigantopithecus blacki 34
glottis *244*
gluteus muscles *67*, 68–9, 70, 71
Goldschmidt, Richard 12
Gongwangling site 181
Goodall, Jane 158, 162, 170
gorillas *(Gorilla gorilla)* 22, *23*; birth 147;
  brain size 118; diet 130; estrus 155–6;
  foramen magnum 74; forest habitat 79,
  168, 259; language 240;
  locomotion 26–7, 77–8; muscles *72*, 81;
  sexual dimorphism 159;
  sociability 158–9, 165; teeth 135, 138;
  testicles 160–1
Gould, Stephen Jay 12, 255
Gracia, Ana 195
Gran Dolina 184–6, 187–90, 192, 210
grasslands 48, 134, 143, 173
Great Rift Valley 57, 173
greenhouse effect 40, 45–6, 262
Greenland 37, 40
gyri 124, *125*, 242

habitat changes 58, 64–5, 87, 172
Hadar site 54, 56–7, 76, 98
Haeckel, Ernst 180
Hahnöfersand fossil 215
Haim, Jean-Louis 248
hamadryas baboons 163, 166
Hammer, Michael 231
hamstring muscles 71, *72*
handaxes 107, *109*
hands 24, 101, 141–2
haplorrhine primates 20–2, 28, 111
haplotypes 231
Harcourt, Alexander 160–1
Henry VIII, King of England 37
herbivores 45–6, 134, 142–3, 186–7
heredity 10; *see also* inheritance
hierarchies 13–14, 22–3
histocompatibility complex 236

Holliday, Trenton 198
Holloway, Ralph 125–6
hominids 15, 59, 76, 112–15;
  bipedalism 53, 54;
  encephalization 123–4; fossils 51, 53,
  57–9, 172–3; hunting 64;
  mandible 54; phonetic
  apparatus 247–8; prey/predators 141;
  teeth 136–8
hominoids 22–7, 33–4, 111; Africa 30–4,
  172; comparative
  sociobiology 158–60; film 64;
  Miocene 30, 47; species 258
Homo spp. *99*; brain size 102, 107,
  113–14, 143–4, 259; emergence 89,
  142; evolutionary diagrams *29, 112, 114,
  189, 210*; jawbones 94, 98; sexual
  dimorphism 102; skulls 102
Homo antecessor 187–90, 210, 218, 224, 235
Homo erectus 102, 179–82, 190, 222–3
Homo ergaster 102, *108*, 114, 181; body
  mass 122, 152; brain 123, 127, 152,
  154, 169, 175; brow ridge 105, *106*;
  megadonty 139; skull 248; social
  change 169; Swartkrans 107
Homo habilis 102, 114; arm–leg
  length 107; body mass 122; brain 123,
  127, 169; brow ridge *103*;
  megadonty 139; Olduvai Gorge 104;
  parietal lobe 242; skull 248
Homo heidelbergensis 209, *210*, 217, 218,
  224
Homo neanderthalensis 216–19
Homo rhodesiensis 224
Homo rudolfensis 98, 102, 104, 114, 123,
  127
Homo sapiens 13, 97–8; African
  origins 223–5; dentition 21–2;
  encephalization 120; Mode 3
  hypothesis 209; and
  Neanderthals 216–19; *see also* humans,
  modern
Howell, Francis Clark 219
Hublin Jean-Jacques 214, 215
humans, modern *23, 25*; airways *245*; birth
  canal 147, *148, 149,* 150; brain
  size 117–20, 128; and chimpanzees 50;
  diet 129–30, 131; DNA 234–5;
  encephalization 128; extensor
  muscles *69; foramen magnum 74;*
  gluteus medius *67*; hamstring
  muscles *72*; jaw/teeth *19*;
  menstruation 146, 165; nature 266–7;
  Neanderthals 212–16; pelvis 70;
  sexuality 155–6; skulls *202*; spine *25,*
  73, 79; *see also Homo sapiens*

humerus 24, 79, 104, 107, 191
hunter-gatherers 131, 164
hunting 64, 101–2, 143
hydrosphere-atmosphere system 40
hyenas 132, 187
Hylobates lar: see gibbons
hyoid bone *245,* 246, 249

Ice Ages 36, 37, 44
Ice Man of Tirol 233
iliac crest 69–71, 73, 146–7, *148*
ilium 69, 71, 81
incisors 134–6, 153–4
India 224
Indian Ocean 44
infants 11, 151–2, 165, 200
information technology 157
inheritance 111, 228–9, 265
intelligence 116, 259–60; adaptation 169;
   carnivores 143–4; climate 64–5;
   Neanderthals 217–18; physical
   features 124, 127–8, 237–8;
   primates 15; technology 264–5
Inuit 131, 200
ischium 69, 71, 150
isolation 238, 255
isotopes 38, 190–1
Israel 185

Java 102, 179–82, 185, 223
jawbones: adolescent 188; *Australopithecus
   afarensis* 58–9; hominoids 34;
   Homo 94; *Paranthropus* 96, 113, 121;
   teeth *133; see also* mandibles; maxilla
Jebel Qafzeh deposit 219, *220,* 223–4
jerboas 53
Johanson, Donald 2, 54, 56, 78, 98, 104
Jungers, William 76, 122, 166

Kanapoi site 137
Kanzi (bonobo) 101, 240–1
Kappelman, John 198
Kebara 221, 249
Keith, Arthur 23–4, 142, 219
Kenya 48, 51
Kenyapithecus 30
Kidd, Judith and Kenneth 232
Kimbel, William 166
Kipling, Rudyard 197
Klasies River Mount 224
Koenigswald, Ralph von 34
Köhler, Meike 32
Konso deposit 91, 105
Koobi Fora 102, 104, 105, 107
Koufos, George 32
Krainitzki, Heike 233

Krakatoa 39
Krapina fossils 211
Krings, Matthias 233
Kromdraai deposit 89, 91
Kubrick, Stanley 64
Kühn, Alfred 263

Laetoli site 54, 56–7, 74–6
Lahr, Marta 209
Laitman, Jeffrey 247, 248, 249
Lamalunga cave 191
Lamarck, Jean-Baptiste de 9–10, 14, 66,
   129, 157
Lana (chimpanzee) 241
Landau, Misia 2
language: articulated 62, 64, 266;
   brain 127, 241–3; chimpanzees 240–1;
   communication 249, 251;
   Lorenz 239–40; Neanderthals 253
laryngeal tone 244–6
larynx 243–4, *245,* 246, *250,* 254
Leakey, Louis 89, 104, 181
Leakey, Mary 53, 74, 89, 104
Leakey, Richard 53, 89, 91, 104, 105
legumes 130, 139, 144
lemurs *20,* 21, 28, 119
Levallois technique 205–6
Levant 217, 219, 221–2
Lezetxiki humerus 191
Lieberman, Philip 247, 249
life expectancy 146, 153–4
life stages 145–6, 152–4
Linné, Karl von 22–3, 97
lions 17, *133*
Lockwood, Charles 166
locomotion 26–7, 64; *see also* bipedalism
Longgupo deposit 181
Lorenz, Konrad 156, 239–40, 263
Lorenzo, Carlos 195
Lorisidae 21, 28
Lothagam deposit 51
Lovejoy, Owen 150, 163–5, 166, *167,*
   168
Lucy: age of fossils 56; bipedalism 1–2, 71,
   76; discovery 54; pelvis 150;
   proportions 78, 79; pubis 151
Lufengpithecus 32

macaques *(Macaca sylvannus)* 18, 161
McCown, Theodore 219
Machado, Antonio 97
McHenry, Henry 122, 138–9, 166
Machiavelli, Niccolò 170
Madagascar 21, 29–30, 119
magnetic field changes 61, 185, 191
Makapansgat site 62

Malawi, Lake 94
Malthus, Thomas Robert 10–11
mammals 14, 87, 118–19
mammoths 233
mandibles: Africa, North 190; Bañolas
   deposit 192; hominids 54; *Homo* 98;
   Kanapoi 137; Lothagam 51;
   Mauer 182, 185, 187, 191, 209–10;
   Paranthropus 94–5, 96
mandibular ramus 96
mandrills 135
Manzi, Giorgio 201, 225
marine paleotemperature scales 38–9, 43,
   190–1
Martínez, Ignacio 187, 195
mass spectrometer 60
masseter muscles *93*, 96
mastication *93*, 94–5, 113, 173, 259; *see
   also* dentition; teeth
Matuyama chron 185, 191
Mauer mandible 182, 185, 187, 191,
   209–10
maxilla *93*, 95, 96, 137, 188, 203
Mayor Cave 192, 194
megadonty, index of 138–9
meiosis 228
menarche 146
Mendel, Gregor 10, 226
meninges 125
menstruation 146, 165; *see also* estrus
Mesozoic 257–8, 260
metabolism 142, 143
meteorite impacts 39, 258
Michelangelo 100
Milankovitch Cycles 40–3
Mimomys savini 185
Miocene 30, 47, 57
missing links 89
mitochondria 265
mitochondrial DNA 229–31, 232, 233–4
Mladec deposit 215
Modjokerto child 179
molars 135–7, 139, 191;
   catarrhines 145–6; chimpanzees 146;
   gorillas 135; life stages 154;
   Paranthropus 153; size 138–41
molecular biology 49
molecular clocks 49–50, 237
monogamy 161–3, 165, *167, 168*
Montaña de Monserrate, Bernardino 239
Morotopithecus 30
Mosquero, Marina 187
mouse lemur 119
Le Mouster 207–8
Mousterian industry 205–6, 212, 214–15,
   217, 221

Moyà-Solà, Salvador 32
multicellular organisms 256–7, 265,
   266
multiregional origin hypothesis 227–8
Mustelidae 134, 192
mutation 12, 50, 261

nails, flat 19, 111
Napier, John 104
nasal bones 105, *106*, 188
nasal cavity 243–4
natural selection: adaptation 10, 238;
   Darwin 10–13, 160, 254;
   evolution 251, 254; genes 156, 164,
   261
Nature 136, 230
Neanderthals 169, 191, 211; brain
   size 201; brow ridge 203; burying the
   dead 206–7, 218; Cro-Magnon
   man 216–19; encephalization 201; facial
   characteristics *193, 202*, 203, *204*, 205,
   208–9; fire 206–7; humans 212–16;
   intelligence 217–18; language 253;
   mitochondrial DNA 233–4;
   neurocranium 201, *202*; phonetic
   apparatus 247; physical
   characteristics 198, *199*, 200, 253;
   skulls 201, *202*; Spain 211; *see also*
   Mousterian industry
neo-Darwinism 12
neocortex 124, 127, 170
Neolithic 130
neurocranium 120–1, 179–80, 201,
   202
newborns: *see* infants
Newton, Sir Isaac 253
Ngandong remains 180, 223
North Pole 40
nose 20–1, 86; *see also* nasal bones
Notarctus *30*
notoungulates 46
nuchal plane 73, 81
nuchal ridge *55*, 81
nuts 131, 141, 144

occipital lobe 124, *125*
occipital ridge 180
Old World monkeys 22, 47, 135, 162,
   172
Oldowan industry 100–2, 107
Olduvai Gorge 104, 107, 181
Oligocene 28
Olorgesailie deposit 48
Omo River 89, 91, 97, 104, 224
Omomyidae 28
oral cavity 243–4, *244*

Index

orangutans *(Pongo pygmaeus)* 22, 23, 34, 259; birth 147; diet 130; females 155–6; hominoid fossils 33; humans 50; locomotion 26; social life 158, 165; temporal muscles 81; testicles 161
Oreopithecus 34
organelles 229, 265
Ouranopithecus 32, 33
Out of Africa hypothesis 226, 236
owl monkey *(Aotus trivirgatus)* 21
oxygen 38–9, 190–1
ozone layer 262

Pääbo, Svante 233
paleoanthropology 4–5, 51, 53, 56, 227
paleocortex 124
paleomagnetism 61
paleomorphology, functional 58, 68
paleoneurology 125, 241–3
paleontology 4, 56, 108–10, 121
Pan paniscus: see chimpanzees, pygmy
Pan troglodytes: see chimpanzees, common
Papua New Guinea 190
Paranthropus 88, 114–15; body mass 122; brain size 143; climate change 175; competition 144; dentition 34, 59, 95, 143; diet 139; emergence 87, 89, 142; environment 162; jawbones 93, 95, 96, 113, 121; life stages 153; mastication 94–5, 173, 259; sagittal ridge 90, 91, 92, 95; sexual dimorphism 173; teeth 139, 140, 153
Paranthropus aethiopicus 89, 90, 173
Paranthropus boisei 47, 89, 173; brain mass 123; skulls 91, 92; teeth size 139, 140
Paranthropus robustus 89, 115, 173; brain mass 123; Swartkrans 91, 107, 141; teeth 139
Parés, Josep María 185
parietal lobe 124, 125, 242
patas monkey *(Erythrocebus patas)* 18, 47
paternity 165
patrilocal societies 232
pelvis 66–7, 69–71; australopithecines 150; bipedalism 24; childbirth 70, 146–7, 148, 149; chimpanzees 68; hyperfeminine 151; La Sima de los Huesos 201; Turkana Boy 152
Pérez-González, Alfredo 185
Petralona deposit 191, 208, 209, 248
phalanges 79, 182
pharynx 243–4, 246
phonetic apparatus 247–8

photosynthesis 14, 45–6
phylogenetic relationships 110, 111
phylum 256
physics 263
Pinilla del Valle deposit 191
Pithecanthropus 179
Pithecanthropus erectus 180
plant types 45–6
platyrrhine primates 21, 29, 30, 255
Plavcan, Michael 166
Pleistocene 47–8, 185; Middle 191, 207; Upper 211
Plesiadapiformes 27, 28
Pliocene 47
pollen grains 39, 207
polymorphism 230, 231, 232
Pongo pygmaeus: see orangutans
porphyria 236
positron emission tomography 242
precipitation 40
premolars 134, 135, 136, 138
Prigogine, Ilya 263
primates 17–22; catarrhine 21–2, 111, 145–6; evolutionary relationships 29; fossils 16, 27; haplorrhine 20–2, 28, 111; history of 27–34; intelligence 15; platyrrhine 21, 29, 255; skeletons 25; strepsirrhine 20–2; teeth 134–6
Proboscidea 46
Proconsul 30, 30–2
prognathism 86, 92, 93, 113, 203
prokaryotes 265, 266
promiscuity 159
protein 130, 131, 132, 139, 143
Protista kingdom 265
Proverbs, Book of 35
puberty 145, 164
pubic bone 200
pubis 70, 150, 151
Purgatorius ceratops 27
Pusey, Anne 170

races 227
radioactive decay 60
rainfall 40, 43
rainforests 18, 172
Rak, Yoel 54
Ramón y Cajal, Santiago ix
recombination 228–9
Reilingen deposit 191, 208
Renne, Grotte de 214, 216
replicants 265
reptiles 257–8, 260
resonance 246
Retzius, striae of 153

rib cages *25,* 79, *82*
Richmond, Brian 166
Robinson, John 91
rodents 30, 185
Rosas, Antonio 187
Rosenberg, Karen 147, 150
Rosny-Aîné, J. H. 216
Ruff, Christopher 152, 198

Saccopastore deposits 211
sacrum 71
Sadiman volcano 74
sagittal ridge *55,* 81, *90,* 91, *92,* 95
Saint-Césaire 214
Sambungmacan braincase 223
Sangiran site 179–80, 181, 223
Sarah (chimpanzee) 241
Savage-Rumbaugh, Sue 101
savanna 43–4, 76, 87, 91, 143, 173
Schaik, Carel van 166
Schmid, Peter 76
Schmitt, Eric-Emmanuel 255
Schmitz, Ralph 233
Schöningen deposit 198
Schrenk, Friedmann 94, 98
Schutz, Adolph 160–1
seafaring 190
seasons 40, *41*
sedimentary deposits 58, 59
seeds 139, 144
selection: artificial 11, 35, 238; group 249,
    251; random 12; sexual 160–2; *see also*
    natural selection
Semendeferi, Katerina 126
Senut, Brigitte 76
sex/love 163–4
sexual dimorphism 56;
    australopithecines 84, 85, 165–6, 167,
    168, 173; chimpanzees 159–60;
    gorillas 159; *Homo* 102, 161;
    Paranthropus 173; La Sima de los
    Huesos 195–6
sexual reproduction 265, 266
sexuality *23,* 155–6
Seyfarth, Robert 240
Shanidar 207
sharks' fins 110
sheep *133*
shoulder blades 24
shrews 118
siamang 24
Silo Cave 192
La Sima de los Huesos 191, 192–6, 201,
    208, 249
Simpson, George Gaylord 217, 256, 264
Sinanthropus 179

Sivapithecus 32–4, 50
Skhul rock shelter 219, 221, 223–4
skulls 81, 209; *Homo* 102, 248;
    Neanderthals 197, 201, *202;*
    neurocranium 120–1; Saccopastore 211;
    Sangiran 179–80; La Sima de los
    Huesos *193,* 249
Smith, Holly 154
Smith, John Maynard 265, 266
snout 20–1
social behavior *23,* 156, 164–5, 169,
    251–2
social biology 78
social skills 127
Socrates 171
solstice 40, *41*
South Africa 94, 173, 224, 236
South Pole 40
Spain 191, 211, 214; *see also individual sites*
spears 198, *199, 213*
specializations 35–6, 94–5, 180, 266
species: emergence 36; evolution 9–10;
    genes 49–50, 216–17;
    groupings 17–18; Linnaean
    classification 22–3;
    paleoanthropology 56
speech organs 243–4, 254
speech production 244–7
speleothems 60, 194
Spielberg, Steven 233
spine 73, *74,* 79
Spoor, Fred 214
Steinheim 191, 208, 248
Sterkfontein site 62, 71, 104
Stern, Jack 76
stomach size 142–4
Stone, Anne 233
stone tools 100–2, 132, 144, 175;
    China 181; Fuente Nueva-3 186; Gran
    Dolina 185; Hadar 98;
    Neanderthals 205–6; *see also*
    Aurignacian; Chatelperronian;
    Mousterian
Stoneking, Mark 230, 233, 237
Story of Life 15, 109, 256, 262, 265
Strassmann, Beverly 165
strepsirrhine primates 20–2
Stringer, Chris 207–8, 209, 226, 248
sulci 124, *125,* 127
Susman, Randall 76
Suwa, Gen 51, 91
Swanscombe deposit 191, 208
Swartkrans 91, 94, 107, 141
Swisher, Carl 179, 223
Szathmáry, Eörs 265, 266
Szeletian industry 213–14

Index

Tabarin site 51
Tabun cave 221
Tague, Robert 150
Tanzania 74, 94, 162–3
taphonomy 58
Tarsiiformes 21
Tattersall, Ian 54–5
Taung Child 61–2, 141, 146, 153
technology 264–5
teeth: apes 135–6; carnassial 132, 134; chimpanzees 138–9, 146; enamel 53, 113, 136–7, 153; herbivores 134; hominids 59, 136–8, 166; mammals 19–20, *133*; *Paranthropus* 139, 140, 153; Pleistocene 191–2; primates 27, 134–6; size 138–41; *see also* canines; dentition; incisors; molars
Teilhard de Chardin, Pierre 17
temporal lobe *117*, 124, *125*
temporalis muscle 81, *93*, 96
Terra Amata 207
Teshik Tash cave 207
testicles 160–1
Therapsida 257–8
thermoluminescence 60–1
Theropithecus spp. 18, *47*, 47–8, *140*, 182
Thieme, Hartmut 198
Thorne, Alan 222–3, 227
Tighenif deposit 190
Tinbergen, Niko 156
Tobias, Phillip 104, 242
toes 24, 111
tongue *245*, 247
tool-making 62, 64, 100–2, 175; bipedalism 77; flakes 212, 213–14; *see also* stone tools
tool-using 65, 98–100
Torres, Trino 192–3
Tossal de la Font de Vilafamés 191
Toth, Nick 101
Trevathan, Wenda 147, 150
Trinchera del Ferrocarril deposits 196
Trinkaus, Erik 198
Trueba, Fernando 5, 264
Truffaut, François 264
tsunami 39
Turkana, Lake 53, 105, 172
Turkana Boy 105, 107, 146, 152–4, 200, 242
Tuttle, Russell 76

'Ubeidiya 185
ulna 79
Uluzzian industry 213–14
Underhill, Peter 231
uranium series dating 60
Ursidae 134

Valdegoba deposit 192
Vandermeersch, Bernard 219
Vega, Gerardo 215
vegetables 131
vegetarian diet 129–30, 142
Vértesszöllös deposit 191, 207, 208
vervet monkeys *(Cercopithecus aethiops)* 240, 241
Victoria, Cueva 182
Vikings 37
Villaverde, Valentín 215
Virgil 212
vision 21
vitalism 14, 16
Viverridae 134
vocal tract 243–4, 244–6
volcanic eruptions 39, 59, 258
volcanic tuffs 59–60
Vrba, Elisabeth 87

Wainscoat, James 232
Walker, Alan 152, 242
Wallace, Alfred Russell 10, 253–4
Washoe (chimpanzee) 241
Weidenreich, Franz 179, 181, 227
Wernicke's area *117*, 241–2
whales 118, 120
Wheeler, Peter 78, 142–4
White, Tim 51, 53, 56, 104, 137, 165, 166
Whiten, Andrew 170
Williams, Jennifer 170
Wilson, Allan 230, 237
Wolpoff, Milford 227
Wood, Bernard 102, 104

Xenophon 155

Y-chromosome 229, 231–2
Yunxian site 224

Zafarraya deposit 215
Zhoukoudian cave 181–2, 207, 209
Zinjanthropus boisei 89
Zuttiyeh cave 221
zygomatic bones *93*, 96, 188, 203